W0079908

Springer Theses

Recognizing Outstanding Ph.D. Research

For further volumes:
http://www.springer.com/series/8790

Aims and Scope

The series "Springer Theses" brings together a selection of the very best Ph.D. theses from around the world and across the physical sciences. Nominated and endorsed by two recognized specialists, each published volume has been selected for its scientific excellence and the high impact of its contents for the pertinent field of research. For greater accessibility to non-specialists, the published versions include an extended introduction, as well as a foreword by the student's supervisor explaining the special relevance of the work for the field. As a whole, the series will provide a valuable resource both for newcomers to the research fields described, and for other scientists seeking detailed background information on special questions. Finally, it provides an accredited documentation of the valuable contributions made by today's younger generation of scientists.

Theses are accepted into the series by invited nominated only and must fulfill all of the following criteria

- They must be written in good English
- The topic of should fall within the confines of Chemistry, Physics and related interdisciplinary fields such as Materials, Nanoscience, Chemical Engineering, Complex Systems and Biophysics.
- The work reported in the thesis must represent a significant scientific advance.
- If the thesis includes previously published material, permission to reproduce this must be gained from the respective copyright holder.
- They must have been examined and passed during the 12 months prior to nomination.
- Each thesis should include a foreword by the supervisor outlining the significance of its content.
- The theses should have a clearly defined structure including an introduction accessible to scientists not expert in that particular field.

Valentin Flunkert

Delay-Coupled Complex Systems

and Applications to Lasers

Doctoral Thesis accepted by
Technische Universität Berlin, Germany

 Springer

Author
Dr. Valentin Flunkert
Institut für Theoretische Physik
Technische Universität Berlin
Hardenbergstrasse 36
10623 Berlin
Germany
e-mail: flunkert@itp.tu-berlin.de

Supervisor
Prof. Dr. Eckehard Schöll, Ph.D.
Institut für Theoretische Physik
Technische Universität Berlin
Hardenbergstrasse 36
10623 Berlin
Germany
e-mail: schoell@physik.tu-berlin.de

ISSN 2190-5053

e-ISSN 2190-5061

ISBN 978-3-642-20249-0

ISBN 978-3-642-20250-6 (eBook)

DOI 10.1007/978-3-642-20250-6

Springer Heidelberg Dordrecht London New York

© Springer-Verlag Berlin Heidelberg 2011

This work is subject to copyright. All rights are reserved, whether the whole or part of the material is concerned, specifically the rights of translation, reprinting, reuse of illustrations, recitation, broadcasting, reproduction on microfilm or in any other way, and storage in data banks. Duplication of this publication or parts thereof is permitted only under the provisions of the German Copyright Law of September 9, 1965, in its current version, and permission for use must always be obtained from Springer. Violations are liable to prosecution under the German Copyright Law.

The use of general descriptive names, registered names, trademarks, etc. in this publication does not imply, even in the absence of a specific statement, that such names are exempt from the relevant protective laws and regulations and therefore free for general use.

Cover design: eStudio Calamar, Berlin/Figueres

Printed on acid-free paper

Springer is part of Springer Science+Business Media (www.springer.com)

To my wife

Supervisor's Foreword

This thesis deals with the effects of time-delay in complex nonlinear systems and in particular with its applications in network and control theory and nonlinear optics. Delays arise naturally in networks of coupled systems due to finite signal propagation speeds and are thus a key issue in many areas of physics, biology, medicine, engineering, economics, and even sociology. Examples are abundant, and include such important applications as lasers, electronic circuits, or neuronal dynamics. Prominent examples from diverse fields are networks of coupled lasers, neuronal networks, communication networks, traffic networks, the internet, power grids, social networks, and many others.

Synchronization phenomena in these networks play an important role, e.g., in the context of learning, cognitive and pathological states in the brain, for secure communication with chaotic lasers or gene regulation. The control of the dynamics on such networks is a fascinating area of research in nonlinear science. A central issue is the interplay of the local dynamics on the nodes and the network topology, the propagation delays, noise, and the role of imperfections. Time-delay extends the dimension of the phase space of a dynamical system to infinity, and thus can dramatically change its stability properties. In particular, unstable periodic or stationary states can be rendered stable, or in contrast, instabilities and bifurcations can be induced. Thus, various states of generalized synchrony, e.g., zero-lag synchronization, antiphase synchronization, or cluster synchronization, can be deliberately selected by tuning the delay-coupling.

This thesis presents novel results on the control of complex dynamics by time-delayed feedback and time-delayed coupling, and new fundamental insights into the interplay of delay and synchronization. As one of the most interesting results this thesis solves the problem of complete synchronization in networks with large coupling delay, i.e., large distances between the nodes, by giving a universal classification of networks, which has a wide range of interdisciplinary applications. The thesis considers in detail the application to lasers, where these results have important consequences with respect to the stability of chaotic synchronization, and bidirectional encrypted communication with a passive or active relay.

Also, the stabilization of periodic intensity pulsations by purely optical or optoelectronic delayed feedback, which can be easily realized experimentally, is investigated.

Berlin, April 2011   Eckehard Schöll

Contents

Notation

For brevity the time argument of dynamic variables will be omitted, when there is no danger of confusion, i.e., we write

$$\frac{d}{dt}x = -\gamma x \quad \text{instead of} \quad \frac{d}{dt}x(t) = -\gamma x(t).$$

In this spirit we denote dynamic variables with delayed arguments in short form by

$$x_\tau = x(t - \tau).$$

Acronyms

ODE	ordinary differential equation
DDE	delay differential equation
RK	Runge–Kutta
LK	Lang–Kobayashi
SM	synchronization manifold
ECM	external cavity mode
LFF	low frequency fluctuation
CC	coherence collapse
UPO	unstable periodic orbit
PO	periodic orbit
FP	fixed point
RW	rotating wave
LE	Lyapunov exponent
TLE	transversal Lyapunov exponent
MSF	master stability function

Part I
Stabilization of Odd-Number Orbits

Chapter 1
Introduction

Delays are ubiquitous in nature and occur, for instance, in coupled systems, in biological processes [1], neural systems [2], or in control problems [3–6]. Time delays arise in these systems due to finite signal propagation and processing speeds, latency effects or are introduced deliberately via external control loops.

From a mathematical point of view, delay terms render a system infinite dimensional (see appendix A) and allow even simple dynamical systems to exhibit complex behavior including oscillations and chaos. The stability of solutions such as periodic orbits and fixed points can change under the influence of delay. This may, on one hand, cause undesired instabilities in engineering applications [7], on the other hand it has led to a completely new research field: The control of dynamical systems by time-delayed feedback.

The control of nonlinear systems is an important topic and has applications in many different fields. In engineering, for instance, nonlinear effects can result in undamped oscillations, which may be undesired or even dangerous. Delayed feedback control is a control scheme that utilizes the system history—in the simplest case the state of the system at an earlier time $t - \tau$, where τ is the delay time—to generate a control signal which is fed back to the system in a closed-loop fashion. The advantages of such a closed-loop control scheme are apparent: there is no need for real time computation of control signals and no reference or target state needs to be known. Instead, the system generates its own control signal and by choosing the parameters of the control loop, e.g., delay time and feedback strengths, appropriately, the system operates in the desired regime.

One particular realization of such a delayed feedback scheme is time-delayed feedback control as proposed by Pyragas [3], which was originally introduced as a method to stabilize unstable periodic orbits but has now found many other applications (see Chap. 2 for a discussion of the control scheme). The most prominent property of the Pyragas control method is its noninvasiveness: If the target orbit is stabilized by the control, then the control force vanishes on the target orbit and therefore the orbit is stabilized but otherwise unchanged. This remarkable feature has drawn a lot of attention to the Pyragas control for two reasons.

V. Flunkert, *Delay-Coupled Complex Systems*, Springer Theses,
DOI: 10.1007/978-3-642-20250-6_1, © Springer-Verlag Berlin Heidelberg 2011

Firstly, the noninvasive stabilization of unstable states makes it possible to study these states in experiments, i.e., unravel dynamical behavior which is usually hidden [5, 8, 9]. Secondly, noninvasive control means that the system is subject to small control signals only. This is important whenever there are limited resources, e.g., constrains due to a finite fuel tank or limits on power consumptions, or when the system to be controlled is fragile and one wants to avoid strong forcing, for example in neural applications.

For ten years it was believed [10] that one of the most common type of periodic orbits, namely odd-number orbits (see Sect. 2.1), could not be stabilized with the Pyragas method. Recently, it was shown that this common believe is in fact wrong and the so-called odd-number theorem was refuted [11]. This surprising turn resulted in a renewed interest in Pyragas control [12–20].

In contrast to delayed feedback control, where the delay is introduced deliberately as a means of control, delays arise naturally in coupled systems due to finite signal propagation speeds. While these latencies may be negligible for small coupling distances, they cannot be ignored when the delay time is comparable the time scales of the dynamical systems. Therefore delays play a crucial role for example in optically coupled lasers [21–26], neuronal [27–29] and biological [30] networks and in dynamical processes in the Internet [31]. Understanding the dynamical behavior of delay coupled systems is thus of great practical importance.

Synchronization phenomena in such networks are relevant [32] in many applications. Chaos synchronization, for instance, may lead to new secure communication schemes [33–35]. The synchronization of neurons is believed to be of great importance in the brain under normal conditions, for instance in the binding problem [36], and under pathological conditions such as Parkinson's disease [37].

The synchronization of delay coupled systems has thus been an important topic in nonlinear science in recent years [24, 25, 38–44].

This thesis is organized in two parts. The first part is devoted to time-delayed feedback control and in particular to the stabilization of odd-number orbits. The second part (p. xx) deals with the synchronization of delay coupled systems. In both parts I consider the application of the results to laser systems. The theoretical and numerical methods used throughout this work are discussed in appendix A. For the efficient simulation of the delay differential equations I have written a simulation package, which is discussed in Chap. 16.

References

1. M.C. Mackey, L. Glass, Oscillation and chaos in physiological control systems. Science **197**, 287 (1977)
2. G. Stepan, Delay effects in brain dynamics. Phil.Trans. R. Soc. A **367**, 1059 (2009)
3. K. Pyragas, Continuous control of chaos by self-controlling feedback. Phys. Lett. A **170**, 421 (1992)
4. K. Pyragas, Delayed feedback control of chaos. Phil. Trans. R. Soc. A **364**, 2309 (2006)

5. E. Schöll, H.G. Schuster (eds.), *Handbook of Chaos Control* (Wiley-VCH, Weinheim, 2008), second completely revised and enlarged edition.
6. W. Just, A. Pelster, M. Schanz, E. Schöll, Delayed complex systems. Theme Issue of Phil. Trans. R. Soc. A **368**, 301–513 (2010)
7. G. Stepan, Modelling nonlinear regenerative effects in metal cutting. Phil. Trans. R. Soc. A **359**, 739 (2001)
8. S. Schikora, P. Hövel, H.J. Wünsche, E. Schöll, F. Henneberger, All-optical noninvasive control of unstable steady states in a semiconductor laser. Phys. Rev. Lett. **97**, 213902 (2006)
9. J. Sieber, A. Gonzalez-Buelga, S. Neild, D. Wagg, B. Krauskopf, Experimental continuation of periodic orbits through a fold. Phys. Rev. Lett. **100**, 244101 (2008)
10. H. Nakajima, On analytical properties of delayed feedback control of chaos. Phys. Lett. A **232**, 207 (1997)
11. B. Fiedler, V. Flunkert, M. Georgi, P. Hövel, E. Schöll, Refuting the odd number limitation of time-delayed feedback control. Phys. Rev. Lett. **98**, 114101 (2007)
12. W. Just, B. Fiedler, V. Flunkert, M. Georgi, P. Hövel, E. Schöll, Beyond odd number limitation: a bifurcation analysis of time-delayed feedback control. Phys. Rev. E **76**, 026210 (2007)
13. B. Fiedler, V. Flunkert, M. Georgi, P. Hövel, E. Schöll, Beyond the odd number limitation of time-delayed feedback control, in *Handbook of Chaos Control*, ed. by E. Schöll, H.G. Schuster (Wiley-VCH, Weinheim, 2008), pp. 73–84, second completely revised and enlarged edition.
14. B. Fiedler, V. Flunkert, M. Georgi, P. Hövel, E. Schöll, Delay stabilization of rotating waves without odd number limitation, in *Reviews of nonlinear dynamics and complexity*, ed. by H.G. Schuster (Wiley-VCH, Weinheim, 2008) vol. **1** pp. 53–68.
15. C.M. Postlethwaite, M. Silber, Stabilizing unstable periodic orbits in the Lorenz equations using time-delayed feedback control. Phys. Rev. E **76**, 056214 (2007)
16. B. Fiedler, S. Yanchuk, V. Flunkert, P. Hövel, H.J. Wünsche, E. Schöll, Delay stabilization of rotating waves near fold bifurcation and application to all-optical control of a semiconductor laser. Phys. Rev. E **77**, 066207 (2008)
17. B. Fiedler: Time-delayed feedback control: Qualitative promise and quantitative constraints, *Proceedings of the 6th EUROMECH Nonlinear Dynamics Conference (ENOC-2008)*, ed. by A. Fradkov, B. Andrievsky (2008), http://lib.physcon.ru/?item=1568
18. M. Kehrt, P. Hövel, V. Flunkert, M.A. Dahlem, P. Rodin, E. Schöll, Stabilization of complex spatio-temporal dynamics near a subcritical Hopf bifurcation by time-delayed feedback. Eur. Phys. J. B **68**, 557 (2009)
19. B. Fiedler, V. Flunkert, P. Hövel, E. Schöll, Delay stabilization of periodic orbits in coupled oscillator systems. Phil. Trans. R. Soc. A **368**, 319 (2010)
20. G. Brown, C.M. Postlethwaite, M. Silber, Time-delayed feedback control of unstable periodic orbits near a subcritical Hopf bifurcation, Physica D (2010), submitted
21. H.J. Wünsche, S. Bauer, J. Kreissl, O. Ushakov, N. Korneyev, F. Henneberger, E. Wille, H. Erzgräber, M. Peil, W. Elsäßer, I. Fischer, Synchronization of delay-coupled oscillators: A study of semiconductor lasers. Phys. Rev. Lett. **94**, 163901 (2005)
22. H. Erzgräber, B. Krauskopf, D. Lenstra, Compound laser modes of mutually delay-coupled lasers. SIAM J. Appl. Dyn. Syst. **5**, 30 (2006)
23. T.W. Carr, I.B. Schwartz, M.Y. Kim, R. Roy, Delayed-mutual coupling dynamics of lasers: scaling laws and resonances. SIAM J. Appl. Dyn. Syst. **5**, 699 (2006)
24. I. Fischer, R. Vicente, J.M. Buldú, M. Peil, C.R. Mirasso, M.C. Torrent, J. García-Ojalvo, Zero-lag long-range synchronization via dynamical relaying. Phys. Rev. Lett. **97**, 123902 (2006)
25. O. D'Huys, R. Vicente, T. Erneux, J. Danckaert, I. Fischer, Synchronization properties of network motifs: Influence of coupling delay and symmetry. Chaos **18**, 037116 (2008)
26. R. Vicente, L.L. Gollo, C.R. Mirasso, I. Fischer, P. Gordon, Dynamical relaying can yield zero time lag neuronal synchrony despite long conduction delays. Proc. Natl. Acad. Sci. **105**, 17157 (2008)

27. E. Rossoni, Y. Chen, M. Ding, J. Feng, Stability of synchronous oscillations in a system of Hodgkin-Huxley neurons with delayed diffusive and pulsed coupling. Phys. Rev. E. **71**, 061904 (2005)

28. C. Hauptmann, O. Omel'chenko, O.V. Popovych, Y. Maistrenko, P.A. Tass, Control of spatially patterned synchrony with multisite delayed feedback. Phys. Rev. E **76**, 066209 (2007)

29. C. Masoller, M.C. Torrent, J. García-Ojalvo, Interplay of subthreshold activity, time-delayed feedback, and noise on neuronal firing patterns. Phys. Rev. E **78**, 041907 (2008)

30. A. Takamatsu, R. Tanaka, H. Yamada, T. Nakagaki, T. Fujii, I. Endo, Spatiotemporal symmetry in rings of coupled biological oscillators of physarum plasmodial slime mold. Phys. Rev. Lett. **87**, 078102 (2001)

31. S.H. Low, F. Paganini, J.C. Doyle, Internet Congestion Control. IEEE Control Systems Magazine **272**, (2002)

32. A.S. Pikovsky, M.G. Rosenblum, J. Kurths, Synchronization, A Universal Concept in Nonlinear Sciences. (Cambridge University Press, Cambridge, 2001)

33. K.M. Cuomo, A.V. Oppenheim, Circuit implementation of synchronized chaos with applications to communications. Phys. Rev. Lett. **71**, 65 (1993)

34. S. Boccaletti, J. Kurths, G. Osipov, D.L. Valladares, C.S. Zhou, The synchronization of chaotic systems. Phys. Rep. **366**, 1 (2002)

35. A. Argyris, D. Syvridis, L. Larger, V. Annovazzi-Lodi, P. Colet, I. Fischer, J. García-Ojalvo, C.R. Mirasso, L. Pesquera, K.A. Shore, Chaos-based communications at high bit rates using commercial fibre-optic links. Nature **438**, 343 (2005)

36. W. Singer, Binding by synchrony. Scholarpedia **2**, 1657 (2007)

37. P.A. Tass, M.G. Rosenblum, J. Weule, J. Kurths, A.S. Pikovsky, J. Volkmann, A. Schnitzler, H.J. Freund, Detection of n:m phase locking from noisy data: Application to magnetoencephalography. Phys. Rev. Lett. **81**, 3291 (1998)

38. A.S. Landsman, L.B. Shaw, I.B. Schwartz. Zero Lag Synchronization of mutually coupled lasers in the presence of delays, in *Recent Advances in Laser Dynamics: Control and Synchronization*, ed. by A.N. Pisarchik (Research Signpost, 2007), p. 359

39. A.S. Landsman, I.B. Schwartz, Complete chaotic synchronization in mutually coupled time-delay systems. Phys. Rev. E **75**, 026201 (2007)

40. V. Flunkert, O. D'Huys, J. Danckaert, I. Fischer, E. Schöll, Bubbling in delay-coupled lasers. Phys. Rev. E **79**, 065201(R) (2009)

41. C.-U. Choe, T. Dahms, P. Hövel, E. Schöll, Controlling synchrony by delay coupling in networks: from in-phase to splay and cluster states. Phys. Rev. E **81**, 025205(R) (2010)

42. M. Zigzag, M. Butkovski, A. Englert, W. Kinzel, I. Kanter, Zero-lag synchronization of chaotic units with time-delayed couplings. Europhys. Lett. **85**, 60005 (2009)

43. W. Kinzel, A. Englert, G. Reents, M. Zigzag, I. Kanter, Synchronization of networks of chaotic units with time-delayed couplings. Phys. Rev. E **79**, 056207 (2009)

44. A. Englert, W. Kinzel, Y. Aviad, M. Butkovski, I. Reidler, M. Zigzag, I. Kanter, M. Rosenbluh, Zero lag synchronization of chaotic systems with time delayed couplings. Phys. Rev. Lett. **104**, 114102 (2010)

Chapter 2
Time-Delayed Feedback Control

Time-delayed feedback control as proposed by Pyragas [1] has proven to be a powerful noninvasive method for the stabilization of unstable periodic orbits (UPOs) [2–5] and unstable steady states [6–9] in dynamical systems. It has since then been successfully applied to a plethora of different systems, for instance, to spatially extended systems [10–16] and even noise-driven systems [17–24], for reviews see [25].

The basic idea is simple, yet the control is often very effective. Consider a dynamical system

$$\frac{d}{dt}X(t) = F(X(t)) \qquad (X \in \mathbb{R}^n)$$

exhibiting an UPO with period T

$$X_*(t) = X_*(t+T).$$

Then the following time-delayed feedback

$$\frac{d}{dt}X(t) = F(X(t)) + K[X(t-\tau) - X(t)],$$

where K is an $n \times n$ real feedback matrix, does not change the orbit if the delay τ is chosen as an integer multiple of the period $\tau = n \cdot T$, since the control force $K[X(t-\tau) - X(t)]$ vanishes on the target orbit. Only the stability properties of the orbit may have changed and for proper choices of the matrix K the formerly unstable orbit may be stabilized. This noninvasive control method can easily be implemented and tested in experimental setups.

We will now discuss a limitation, which was thought to exist and which we will refute in the following chapters.

V. Flunkert, *Delay-Coupled Complex Systems*, Springer Theses,
DOI: 10.1007/978-3-642-20250-6_2, © Springer-Verlag Berlin Heidelberg 2011

2.1 Alleged Odd-Number Theorem

Severe restrictions for the applicability of the method were believed to exist
[26–31]. It was commonly believed that UPOs with an odd-number of real
Floquet multipliers larger than unity could never be stabilized by delayed
feedback control (see Sect. 15.2 for a definition of Floquet exponents and
multipliers). Note that many of the most commonly found UPOs in dynamical
systems belong to this class of odd-number orbits. For instance, any UPO born
in a subcritical Hopf bifurcation from an unstable fixed point (FP) as well as
any UPO born in a saddle-node bifurcation of periodic orbits (POs) with a
stable partner is an odd-number orbit.

Recently this alleged *odd-number theorem* has been refuted by counterexam-
ples [32–37], which I will present below. The proof of the odd-number theorem
provided in [27] fails because it does not take the Goldstone mode, i.e, the trivial
Floquet multiplier $\mu = 1$, of POs in autonomous systems into account. The odd-
number theorem actually remains valid for UPOs, which do not have a trivial
Floquet multiplier, i.e., orbits, which are induced by external time-dependent
forcing. However, such orbits occur less frequently in practice. For a detailed
discussion of the odd-number theorem's proof and its shortcoming see [38]. We
will now provide a counterexample to the odd-number theorem.

References

1. K. Pyragas, Continuous control of chaos by self-controlling feedback. Phys. Lett. A **170**, 421 (1992)
2. E. Schöll, K. Pyragas, Tunable semiconductor oscillator based on self-control of chaos in the dynamic Hall effect. Europhys. Lett. **24**, 159 (1993)
3. O. Beck, A. Amann, E. Schöll, J.E.S. Socolar, W. Just, Comparison of time-delayed feedback schemes for spatio-temporal control of chaos in a reaction-diffusion system with global coupling. Phys. Rev. E **66**, 016213 (2002)
4. N. Baba, A. Amann, E. Schöll, W. Just, Giant improvement of time-delayed feedback control by spatio-temporal filtering. Phys. Rev. Lett. **89**, 074101 (2002)
5. C. von Loewenich, H. Benner, W. Just, Experimental relevance of global properties of time-delayed feedback control. Phys. Rev. Lett. **93**, 174101 (2004)
6. P. Hövel, E. Schöll, Control of unstable steady states by time-delayed feedback methods. Phys. Rev. E **72**, 046203 (2005)
7. S. Schikora, P. Hövel, H.J. Wünsche, E. Schöll, F. Henneberger, All-optical noninvasive control of unstable steady states in a semiconductor laser. Phys. Rev. Lett. **97**, 213902 (2006)
8. T. Dahms, P. Hövel, E. Schöll, Control of unstable steady states by extended time-delayed feedback. Phys. Rev. E **76**, 056201 (2007)
9. T. Dahms, P. Hövel, E. Schöll, Stabilizing continuous-wave output in semiconductor lasers by time-delayed feedback. Phys. Rev. E **78**, 056213 (2008)
10. G. Franceschini, S. Bose, E. Schöll, Control of chaotic spatiotemporal spiking by time-delay autosynchronisation. Phys. Rev. E **60**, 5426 (1999)
11. J. Unkelbach, A. Amann, W. Just, E. Schöll, Time–delay autosynchronization of the spatiotemporal dynamics in resonant tunneling diodes. Phys. Rev. E **68**, 026204 (2003)

12. J. Schlesner, A. Amann, N.B. Janson, W. Just, E. Schöll, Self-stabilization of high frequency oscillations in semiconductor superlattices by time–delay autosynchronization. Phys. Rev. E **68**, 066208 (2003)
13. M.A. Dahlem, F.M. Schneider, E. Schöll, Failure of feedback as a putative common mechanism of spreading depolarizations in migraine and stroke. Chaos **18**, 026110 (2008)
14. M.A. Dahlem, R. Graf, A.J. Strong, J.P. Dreier, Y.A. Dahlem, M. Sieber, W. Hanke, K. Podoll, E. Schöll, Two-dimensional wave patterns of spreading depolarization: retracting, re-entrant, and stationary waves. Physica D **239**, 889 (2010)
15. F.M. Schneider, E. Schöll, M.A. Dahlem, Controlling the onset of traveling pulses in excitable media by nonlocal spatial coupling and time delayed feedback. Chaos **19**, 015110 (2009)
16. Y.N. Kyrychko, K.B. Blyuss, S.J. Hogan, E. Schöll, Control of spatio-temporal patterns in the Gray-Scott model. Chaos **19**, 043126 (2009)
17. N.B. Janson, A.G. Balanov, E. Schöll, Delayed feedback as a means of control of noise-induced motion. Phys. Rev. Lett. **93**, 010601 (2004)
18. J. Pomplun, A. Amann, E. Schöll, Mean field approximation of time-delayed feedback control of noise-induced oscillations in the Van der Pol system. Europhys. Lett. **71**, 366 (2005)
19. G. Stegemann, A.G. Balanov, E. Schöll, Delayed feedback control of stochastic spatiotemporal dynamics in a resonant tunneling diode. Phys. Rev. E **73**, 016203 (2006)
20. V. Flunkert, E. Schöll, Suppressing noise-induced intensity pulsations in semiconductor lasers by means of time-delayed feedback. Phys. Rev. E **76**, 066202 (2007)
21. T. Prager, H.P. Lerch, L. Schimansky-Geier, E. Schöll, Increase of coherence in excitable systems by delayed feedback. J. Phys. A **40**, 11045 (2007)
22. A. Pototsky, N.B. Janson, Correlation theory of delayed feedback in stochastic systems below andronov-hopf bifurcation. Phys. Rev. E **76**, 056208 (2007)
23. J. Hizanidis, E. Schöll, Control of coherence resonance in semiconductor superlattices. Phys. Rev. E **78**, 066205 (2008)
24. N. Majer, E. Schöll, Resonant control of stochastic spatio-temporal dynamics in a tunnel diode by multiple time delayed feedback. Phys. Rev. E **79**, 011109 (2009)
25. E. Schöll, in *Handbook of Chaos Control*, ed. by H.G. Schuster (Wiley-VCH, Weinheim, 2008), second completely revised and enlarged edition.
26. W. Just, T. Bernard, M. Ostheimer, E. Reibold, H. Benner, Mechanism of time-delayed feedback control. Phys. Rev. Lett. **78**, 203 (1997)
27. H. Nakajima, On analytical properties of delayed feedback control of chaos. Phys. Lett. A **232**, 207 (1997)
28. H. Nakajima, Y. Ueda, Limitation of generalized delayed feedback control. Physica D **111**, 143 (1998)
29. I. Harrington, J.E.S. Socolar, Limitation on stabilizing plane waves via time-delay feedback. Phys. Rev. E **64**, 056206 (2001)
30. K. Pyragas, V. Pyragas, H. Benner, Delayed feedback control of dynamical systems at subcritical Hopf bifurcation. Phys. Rev. E **70**, 056222 (2004)
31. V. Pyragas, K. Pyragas, Delayed feedback control of the Lorenz system: An analytical treatment at a subcritical Hopf bifurcation. Phys. Rev. E **73**, 036215 (2006)
32. B. Fiedler, V. Flunkert, M. Georgi, P. Hövel, E. Schöll, Refuting the odd number limitation of time-delayed feedback control. Phys. Rev. Lett. **98**, 114101 (2007)
33. B. Fiedler, V. Flunkert, M. Georgi, P. Hövel, E. Schöll (2008) in *Beyond the odd number limitation of time-delayed feedback control*, ed. by Schöll E., Schuster H.G. Handbook of Chaos Control, Wiley-VCH, Weinheim, 73–84, second completely revised and enlarged edition
34. B. Fiedler, V. Flunkert, M. Georgi, P. Hövel, E. Schöll, in Delay stabilization of rotating waves without odd number limitation, ed. by H.G. Schuster Reviews of nonlinear dynamics and complexity, vol. 1, (Wiley VCH, Weinheim, 2008) pp. 53–68

35. W. Just, B. Fiedler, V. Flunkert, M. Georgi, P. Hövel, E. Schöll, Beyond odd number limitation: a bifurcation analysis of time-delayed feedback control. Phys. Rev. E **76**, 026210 (2007)
36. B. Fiedler, S. Yanchuk, V. Flunkert, P. Hövel, H.J. Wünsche, E. Schöll, Delay stabilization of rotating waves near fold bifurcation and application to all-optical control of a semiconductor lase. Phys. Rev. E **77**, 066207 (2008)
37. M. Kehrt, P. Hövel, V. Flunkert, M.A. Dahlem, P. Rodin, E. Schöll, Stabilization of complex spatio-temporal dynamics near a subcritical Hopf bifurcation by time-delayed feedback. Eur. Phys. J. B **68**, 557 (2009)
38. P. Hövel (2009) Control of complex nonlinear systems with delay, Ph.D. thesis, Technische Universität Berlin

Chapter 3
Counterexample

In this section we will construct a counterexample to the odd-number theorem, i.e., a system with an odd-number orbit, where the orbit can be stabilized by time-delayed feedback control. The counterexample consists of the normal form of a subcritical Hopf bifurcation

$$\frac{d}{dt}z = \left[\lambda + i + (1 + i\gamma)|z|^2\right]z \qquad (z \in \mathbb{C}), \tag{3.1}$$

where the time is scaled such that the frequency is one $(\omega = 1)$. Written in amplitude and phase $z(t) = r(t)e^{i\theta(t)}$ the equation becomes

$$\frac{d}{dt}r = (\lambda + r^2)r, \tag{3.2a}$$

$$\frac{d}{dt}\theta = 1 + \gamma r^2. \tag{3.2b}$$

For $\lambda < 0$ an UPO with $r = \sqrt{-\lambda}$ and period $T = 2\pi/(1 - \gamma\lambda)$ exists. At $\lambda = 0$ a subcritical Hopf bifurcation occurs and the FP $z = 0$ becomes unstable for $\lambda > 0$. Figure 3.1 depicts the bifurcation diagram (panel (a)) and the period (panel (b) and (c)) of the UPO. The PO is born at $\lambda = 0$ with a finite period of $T = 2\pi$, which then increases or decreases for decreasing λ depending on the sign of γ.

Equation (3.1) describes an autonomous system, and thus one of the orbit's Floquet multipliers is unity, corresponding to the Goldstone mode, i.e., the phase shift invariance of the orbit. Since the orbit is unstable, the other Floquet multiplier is larger than one and the orbit is an UPO belonging to the odd-number class. This is the target orbit we wish to stabilize. We will call this orbit the *Pyragas orbit*

Following [1] we will now stabilize the Pyragas orbit by applying time-delayed feedback control

$$\frac{d}{dt}z = [\lambda + i + (1 + i\gamma)|z|^2]z + b[z_\tau - z]. \tag{3.3}$$

V. Flunkert, *Delay-Coupled Complex Systems*, Springer Theses,
DOI: 10.1007/978-3-642-20250-6_3, © Springer-Verlag Berlin Heidelberg 2011

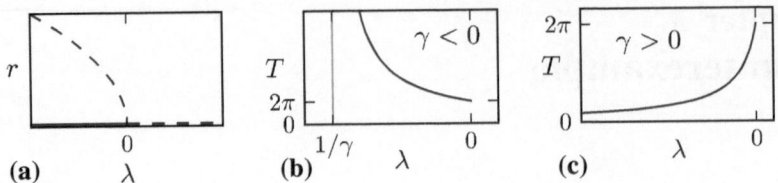

Fig. 3.1 Bifurcation diagram for the subcritical Hopf bifurcation. Panel (**a**): Radius of the UPO and the FP at $z = 0$. *Solid* and *dashed curves* correspond to stable and unstable solutions, respectively. Panel (**b**) and (**c**): Period T of the UPO for $\gamma < 0$ and $\gamma > 0$

Here, z_τ denotes the time-delayed variable (see notation on p. xi) and $b = b_0 e^{i\beta}$ is a complex feedback gain.

We could go ahead and try to analyze the PO's stability using Floquet theory and try to find successful control forces b. However, the Floquet problem for delay differential equations (DDEs) leads in this case to a nonlinear transcendental equation (see Sect. 3.3) and is very difficult to treat analytically.

The way this problem can be approached, nevertheless, is to construct conditions such that the FP $z = 0$ is unstable for $\lambda < 0$ and stable for $\lambda > 0$. If one succeeds while preserving the location of the PO, as is the case for noninvasive time-delayed feedback control, then the subcritical Hopf bifurcation must have become supercritical and the PO will be stable at least in the vicinity of $\lambda = 0$.

For noninvasive control the value of τ is determined by the period of the PO, i.e., the delay time has to be chosen as $\tau = n \cdot T$. For each $n \in \mathbb{N}$ this defines curves in the (λ, τ)-plane, which we call the *Pyragas curve*

$$\tau_p(\lambda) = \frac{2\pi n}{1 - \gamma \lambda} \qquad (\lambda < 0). \tag{3.4}$$

The Pyragas curves emanate from the points $(\lambda, \tau) = (0, 2\pi n)$ and extend to the left half-plane $(\lambda < 0)$. For negative and positive γ the curve goes up and down (see Fig. 3.1), respectively. We are restricted to this curve in the (λ, τ)-plane for given a γ.

To address the stability of the FP, we linearize (3.3) around the FP $z = 0$

$$\frac{d}{dt} z = (\lambda + i)z + b[z_\tau - z].$$

Making the ansatz $z(t) \propto e^{\eta t}$ we obtain a transcendental characteristic equation for the complex eigenvalues $\eta \in \mathbb{C}$, which govern the FP's stability

$$\eta = \lambda + i + b(e^{-\eta \tau} - 1). \tag{3.5}$$

To find bifurcation lines in the parameter plane, i.e., the boundary of the FP's stability domain, we seek solutions with $\text{Re}(\eta) = 0$. Since $\eta = 0$ is not a solution of the characteristic equation, one has to look for solutions $\eta = i\Omega$ corresponding to Hopf bifurcations. Inserting this ansatz into (3.5) and splitting the equation into real and imaginary part we find

Fig. 3.2 Hopf curves in the (λ, τ)-plane. The *curves* correspond with increasing darkness to $b_0 = 0$, 0.02, 0.1, 0.2, 0.3. Parameter: $\beta = \pi/4$

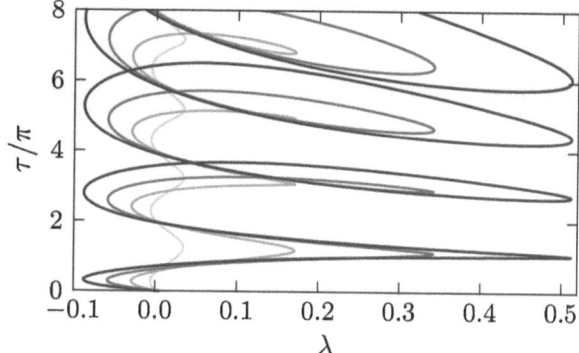

$$0 = \lambda + b_0[\cos(\beta - \Omega\tau) - \cos\beta], \qquad (3.6a)$$

$$\Omega - 1 = b_0[\sin(\beta - \Omega\tau) - \sin\beta]. \qquad (3.6b)$$

Using some trigonometric identities we can eliminate Ω to obtain τ as a function of λ. These *Hopf curves* $\tau_H(\lambda)$ are given by

$$\tau_H(\lambda) = \frac{\pm \arccos\left(\frac{b_0 \cos\beta - \lambda}{b_0}\right) + \beta + 2\pi n}{1 - b_0 \sin\beta \mp \sqrt{\lambda(2b_0 \cos\beta - \lambda) + b_0^2 \sin^2\beta}}, \qquad (3.7)$$

where the upper and lower signs correspond to different branches. Note that τ_H is not defined for $\beta = 0$ and $\lambda < 0$. Since we want to destabilize the FP for $\lambda < 0$, it is necessary to choose complex b.

Figure 3.2 depicts the family of Hopf curves for fixed $\beta = \pi/4$ and different b_0. The line $\lambda = 0$ corresponds to the uncontrolled system ($b_0 = 0$), which has a Hopf bifurcation at $\lambda = 0$ independent of the value of τ. With increasing b_0 the Hopf curves stretch further into the $\lambda < 0$ half-plane. Note that all Hopf curves pass through the points ($\lambda = 0$, $\tau = 2\pi n$). This is due to the fact that $\lambda = 0$, $\Omega = 1$, and $\tau = 2\pi n$ is always a solution of (3.5).

Figure 3.3 depicts the Hopf curve for $b_0 = 0.3$ and $\beta = \pi/4$. The numbers in parentheses and the shading indicate the unstable dimensions of the FP, i.e., (0)– stable, (2)– two-fold unstable etc. To destabilize the FP $z = 0$ for $\lambda < 0$ and thereby stabilizing the PO we have to choose points (λ, τ) on the Pyragas curve within the light blue region, where the FP is two-fold unstable.

This means we have to find values of $b = b_0 e^{i\beta}$, for which the Hopf curves are organized such that the Pyragas curve reaches into the light blue region, where the FP is two-fold unstable. In this case when moving along the Pyragas curve the Hopf bifurcation has changed from subcritical to supercritical and thus the PO is stabilized. This desired situation is shown in Fig. 3.4. The Pyragas curve extends into the region where the FP is two-fold unstable. Along the Pyragas curve, with

Fig. 3.3 Hopf curves and unstable dimensions (*in brackets*) of the FP in the (λ, τ)-plane. Parameters: $b_0 = 0.3$, $\beta = \pi/4$

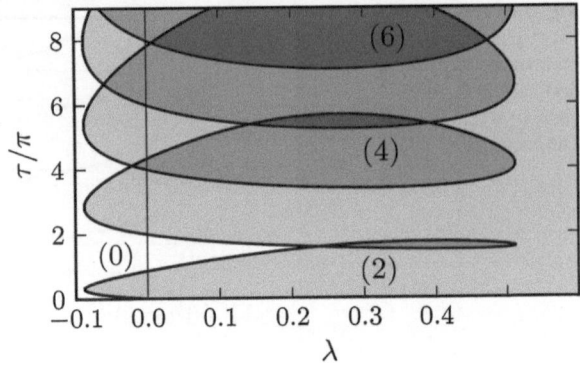

Fig. 3.4 Pyragas curve (*dashed*) and Hopf curves in the (λ, τ)-plane corresponding to (3.4) and (3.7), respectively. Numbers in parentheses denote the unstable dimension of the FP. Parameters: $b_0 = 0.3$, $\beta = \pi/4$, $\gamma = -10$

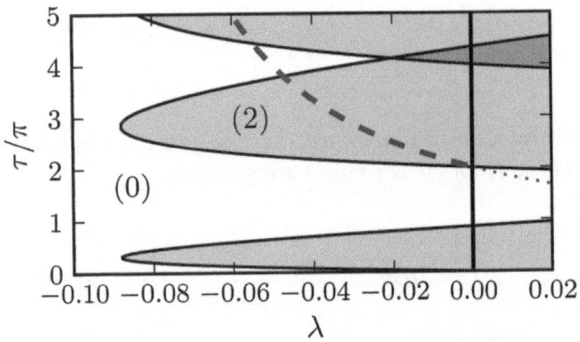

the dotted extension, the Hopf bifurcation has changed from subcritical to supercritical.

The sign of γ determines whether the Pyragas curve points upwards $(\gamma < 0)$ or downwards $(\gamma > 0)$ (see Fig. 3.1). These two cases have to be treated separately. As it turns out, switching the sign of β locally reflects the Hopf curve along the $\tau = 2\pi$ line, which will allow us to stabilize either case. To see this we can insert $\Omega = 1 + \delta\Omega$ and $\tau = 2\pi + \delta\tau$ into (3.7). Then in linear order the resulting equations are invariant under the transformation $\beta \to -\beta$, $\delta\tau \to -\delta\tau$ and $\delta\Omega \to -\delta\Omega$. This symmetry is depicted in Fig. 3.5. Switching the sign of β approximately reflects the Hopf curve along the $\tau = 2\pi$ line (and simultaneously along any of the lines $\tau = 2\pi n$). Note that for $\beta = 0$ the Hopf curve does not reach into the $\lambda < 0$ half-plane.

We will now construct the stabilization conditions for the two cases $\gamma < 0$ and $\gamma > 0$, which are called soft spring and hard spring case, respectively. The mechanical terminology "soft" and "hard" spring arises from the pendulum equation $\ddot{x} + Dx = 0$ with nonlinear spring constant $D = D(x)$. For "soft" springs $D(x)$, where $D(x)$ decreases with increasing $|x|$, the period increases with increasing amplitude. Examples are mathematical pendula $D(x) = \sin x$ or rubber balloons. For "hard" springs $D(x)$ increases with $|x|$ and the period decreases with increasing amplitude.

Fig. 3.5 Hopf curve for different values of β. Parameter: $b_0 = 0.05$

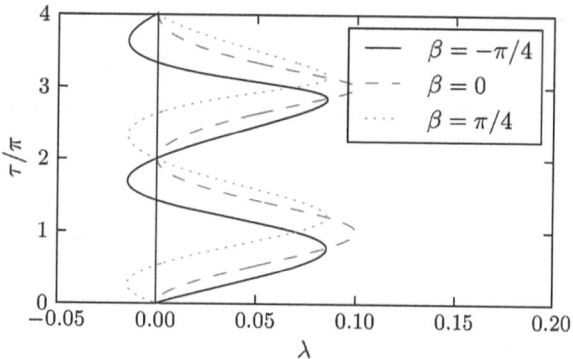

Consider the τ-axis in the (λ, τ)-plane. The intersections of the Hopf curve with this axis are obtained from (3.6) by setting $\lambda = 0$

$$\cos \beta = \cos(\beta - \Omega\tau),$$
$$\frac{\Omega - 1}{b_0} + \sin \beta = \sin(\beta - \Omega\tau).$$

Squaring both equations and adding them eliminates τ and results in a quadratic equation for Ω

$$\Omega^2 - (b_0 \sin \beta - 2)\Omega + 1 - b_0 \sin \beta = 0.$$

The two roots lead to two sets of solutions

$$\tau_n^A = 2\pi n, \qquad \Omega^A = 1, \tag{3.8a}$$

$$\tau_n^B = \frac{2\beta + 2\pi n}{1 - 2b_0 \sin \beta}, \quad \Omega^B = 1 - 2b_0 \sin \beta, \tag{3.8b}$$

with $n = 0, 1, 2, \ldots$. Note that the A series consists of all points, where the Pyragas curves emanate.

To calculate the unstable dimensions of the FP above and below each Hopf point we will now analyze the crossing direction of the Hopf eigenvalues with increasing τ. The eigenvalue equation on the τ-axis is given by

$$\eta = i + b(e^{-\eta\tau} - 1).$$

Implicit differentiation with respect to τ gives

$$\partial_\tau \eta = b(-\eta - \tau\partial_\tau \eta)e^{-\eta\tau}. \tag{3.9}$$

The crossing direction of the Hopf eigenvalues is determined by the sign of $\mathrm{Re}(\partial_\tau\eta)$. For $\mathrm{Re}(\partial_\tau\eta) > 0$ the unstable dimensions of the FP increase by two when going up the τ-axis through the Hopf points, and for $\mathrm{Re}(\partial_\tau\eta) < 0$ it is vice-versa. Evaluating $\mathrm{Re}(\partial_\tau\eta)$ at the Hopf points ($\eta = i\Omega^{A,B}$, $\tau = \tau^{A,B}$) yields

$$\mathrm{Re}(\partial_\tau \eta) = \mathrm{Re}\left(-\frac{b\eta}{1 + \tau e^{-\eta\tau}}\right) = \begin{cases} \sin\beta & \text{for the A series,} \\ \sin\beta(2b_0 \sin\beta - 1) & \text{for the B series.} \end{cases}$$

We now have all ingredients to find conditions, which allow stabilization: At the Hopf points of series A, where the Pyragas curve emanates, we can decide whether the Hopf point lies on the border between a (0)-region and a (2)-region and which region lies on which side. From the explicit forms of the Hopf and the Pyragas curve, we can decide whether locally the Pyragas curve reaches into the (2)-region.

3.1 The Case $\gamma < 0$

For $\gamma < 0$ the Pyragas curve points upwards $(\tau'_P(0) < 0)$ from the emanating point into the left half-plane. It is therefore necessary that there is a (2)-region above the Hopf point and a (0)-region below.

In order to have a change from unstable dimension (0) to (2) of the FP, the emanating point, which belongs to the A series, needs $\mathrm{Re}(\partial_\tau \eta) = \sin\beta > 0$ and thus

$$0 < \beta < \pi \tag{3.10}$$

or equivalently $\mathrm{Im}(b) > 0$.

To have a (0)-region below the emanating point the Hopf curves have to turn back across the τ-axis. Firstly, this requires that the B points have $\mathrm{Re}(\eta\tau) < 0$, i.e.,

$$0 < \beta < \arcsin\left(\frac{1}{2b_0}\right) \quad \text{or} \quad \pi - \arcsin\left(\frac{1}{2b_0}\right) < \beta < \pi. \tag{3.11}$$

Secondly, every A point below the emanating point has to be compensated by a B point, which implies the following ordering

$$\tau_0^A < \tau_0^B < \tau_1^A < \tau_1^B < \cdots < \tau_{n-1}^B < \tau_n^A. \tag{3.12}$$

The distance between two successive B points is given by (see (3.8))

$$\tau_{k+1}^B - \tau_k^B = \frac{2\pi}{1 - 2b_0 \sin\beta} > 2\pi,$$

where we used that $\tau_k^B > 0$, implying $1 > 2b_0 \sin\beta$. Since the distance between two successive A points is 2π, there is at most one B point between two A points. With increasing $\beta \in [0, \pi]$ the distance between two successive B points becomes larger. The order of (3.12) is first violated when $\tau_{n-1}^B = \tau_n^A$ for some $\beta = \beta_n^-$. Inserting τ_{n-1}^B and τ_n^A and solving for β_n^- gives a transcendental equation

$$\frac{1}{\pi}\beta_n^- + 2nb_0 \sin\beta_n^- = 1. \tag{3.13}$$

This yields the first stabilization condition

$$0 < \beta < \beta_n^-. \tag{3.14}$$

We have now established conditions, which lead to a (0)-region below the emanating point and a (2)-region above. The last requirement is that the Pyragas curve reaches into the (2)-region, i.e., the Pyragas curve runs above the Hopf curve, locally. This is satisfied if

$$\partial_\lambda \tau_P < \partial_\lambda \tau_H \quad \text{at} \quad (\lambda, \tau) = (0, 2\pi n). \tag{3.15}$$

From (3.4) we find

$$\partial_\lambda \tau_P \Big|_{(\lambda, \tau) = (0, 2\pi n)} = 2\pi n \gamma.$$

Implicit differentiation of (3.6a, 3.6b) with respect to λ gives

$$0 = 1 + (\Omega \partial_\lambda \tau + \tau \partial_\lambda \Omega) b_0 \sin(\beta - \Omega \tau)$$
$$\partial_\lambda \Omega = -(\Omega \partial_\lambda \tau + \tau \partial_\lambda \Omega) b_0 \cos(\beta - \Omega \tau).$$

Inserting $\lambda = 0$, $\tau = 2\pi n$ and eliminating $\partial_\lambda \Omega$ gives the desired slope

$$\partial_\lambda \tau_H \Big|_{(\lambda, \tau) = (0, 2\pi n)} = -\frac{1 + 2\pi n b_0 \cos \beta}{b_0 \sin \beta} = -\frac{1 + 2\pi n \mathrm{Re}(b)}{\mathrm{Im}(b)}.$$

Stabilization is therefore possible if

$$\frac{1}{\mathrm{Im}(b)} \left(\mathrm{Re}(b) + \frac{1}{2\pi n} \right) < -\gamma. \tag{3.16}$$

To summarize we have found two conditions (3.14) and (3.16) on the feedback constant b, which imply stabilization of the UPO close to the bifurcation $\lambda = 0$. The domain of control is thus bounded by the two curves

$$b_0(\beta) = \frac{1}{2n \sin \beta} \left(1 - \frac{\beta}{\pi} \right), \tag{3.17a}$$

$$\mathrm{Im}(b) = -\frac{1}{\gamma} \left(\mathrm{Re}(b) + \frac{1}{2\pi n} \right), \tag{3.17b}$$

where we solved (3.13) for b_0 as a function of β. The domain of control resulting from the two curves is depicted in Fig. 3.6(a) for fixed n and different values of γ and Fig. 3.6(b) for fixed γ and different values of n.

To illustrate the stabilization, we simulated the equations with the control force b chosen in the control domain. Figure 3.7 depicts the time series of the amplitude $|z|$ and $|z - z_\tau|$, which is proportional to the control signal. The system starts with the constant history $z(t) = 0.1$ for $t \in [-\tau, 0]$, which is not very close to the Pyragas orbit. After some transient time, the system approaches the Pyragas orbit

Fig. 3.6 Domains of control in the plane of complex feedback gain $b = b_0 e^{i\beta}$ for different values of n and γ in the limit $\lambda \nearrow 0$. Panel (**a**): The *shaded regions* correspond to $\gamma = -10, -5, -2$ with increasing darkness ($n = 1$). Panel (**b**): The *shaded regions* correspond to $n = 1, 2, 3$ with increasing darkness ($\gamma = -10$)

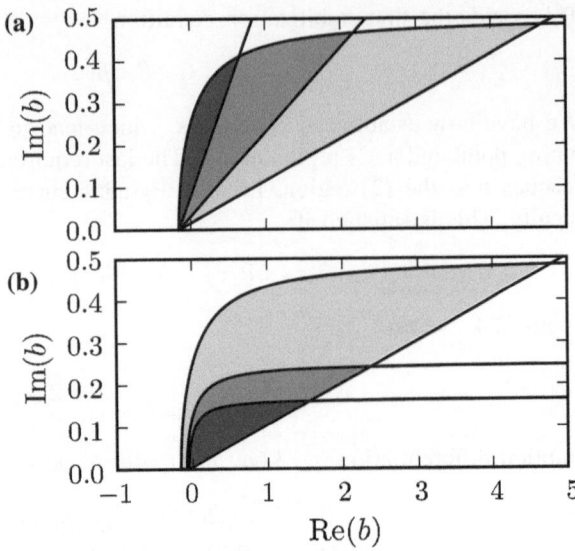

Fig. 3.7 Time series in the case of stabilization. Panel (**a**): Time series of the absolute value $|z|$. Panel (**b**): semi-log plot of $|z - z_\tau|$ vs time t. Parameters: $\lambda = -0.005, \gamma = -10$, $b = 0.3 e^{i\pi/4}$, $\tau = 2\pi/(1 - \gamma\lambda)$

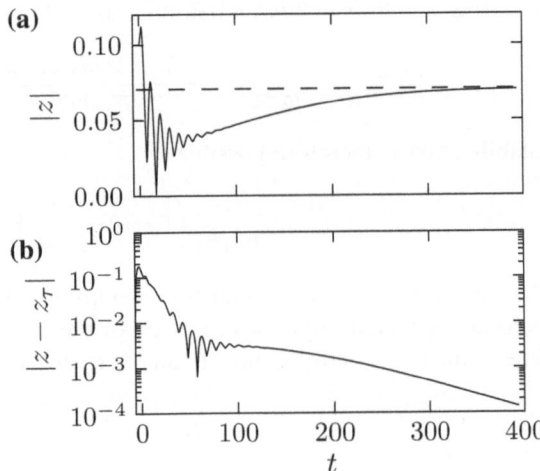

indicated by the dashed line. The closer the trajectory comes to the Pyragas orbit, the smaller the control signal becomes, due to the noninvasive nature of the Pyragas method.

3.2 The Case $\gamma > 0$

For $\gamma > 0$ the Pyragas curve has a positive slope at $\lambda = 0$ and thus reaches downwards from the emanating point into the left half-plane $\lambda < 0$ (see

Fig. 3.1(c)). In this case it is necessary that the Hopf curve, which passes through the emanating point, separates a (0)-region above from a (2)-region below, i.e., $\mathrm{Re}(\partial_\tau \eta) = \sin \beta < 0$ and thus

$$\pi < \beta < 2\pi. \tag{3.18}$$

Additionally, the B point of the closing Hopf curve above the emanating point is not allow to lie below the emanating point. Otherwise there would be a $(2) \to (4)$ change at the emanating point with decreasing τ instead of a $(0) \to (2)$ change. This implies the following ordering of Hopf points along the τ-axis

$$\tau_0^A < \tau_1^B < \tau_1^A < \cdots < \tau_n^A < \tau_{n+1}^B. \tag{3.19}$$

Note the slight difference to the ordering (3.12) in the case of $\gamma < 0$. With the same argumentation as above this ordering is first violated when $\tau_n^A = \tau_{n+1}^B$ for some $\beta = \beta_n^+$. Inserting τ_n^A and τ_{n+1}^B results in the transcendental equation for β_n^+, which differs from (3.13) by a minus sign

$$\frac{1}{\pi} \beta_n^+ + 2nb_0 \sin \beta_n^+ = -1. \tag{3.20}$$

This again gives a second stabilization condition

$$\beta_n^+ < \beta < 2\pi. \tag{3.21}$$

In order for the Pyragas curve to reach into the (2)-region, we also need

$$\partial_\lambda \tau_P > \partial_\lambda \tau_H \quad \text{at} \quad (\lambda, \tau) = (0, 2\pi n),$$

which results in

$$\frac{1}{\mathrm{Im}(b)} \left(\mathrm{Re}(b) + \frac{1}{2\pi n} \right) > -\gamma. \tag{3.22}$$

Together with (3.21) this gives the boundary of the control domain

$$b_0(\beta) = -\frac{1}{2n \sin \beta} \left(1 + \frac{\beta}{\pi} \right), \tag{3.23a}$$

$$\mathrm{Im}(b) = -\frac{1}{\gamma} \left(\mathrm{Re}(b) + \frac{1}{2\pi n} \right). \tag{3.23b}$$

Figure 3.8 depicts the control domain in the complex b-plane for different values of $\gamma > 0$ and n. Compared with Fig. 3.6 the control is reflected along the $\mathrm{Im}(b) = 0$ axis. This can be seen by transforming (3.23) according to $\beta \to -\beta$ and $\gamma \to -\gamma$, which yields (3.17). In the following we will restrict the analysis to $\gamma < 0$ since all results can be reproduced in a similar manner for $\gamma > 0$.

Having found parameter domains, where the FP is unstable, we will now numerically investigate the stability of the Pyragas orbit using Floquet theory for

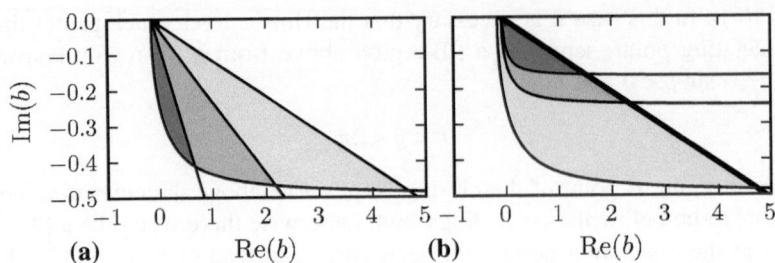

Fig. 3.8 Domains of control in the plane of complex feedback gain $b = b_0 e^{i\beta}$ for different values of n and γ. Panel (**a**): The *shaded regions* correspond to $\gamma = 10, 5, 2$ with increasing darkness ($n = 1$). Panel (**b**): The *shaded regions* correspond to $n = 1, 2, 3$ with increasing darkness ($\gamma = 10$)

the delayed system. In general this would only be possible by using tools for numerical bifurcation analysis of delay equations, such as DDE-BIFTOOL [2, 3] or KNUT (formerly known as PDDE-CONT) [4]. In our example, however, the POs respect the rotation symmetry of the system, which allows a semi-analytic treatment.

3.3 Floquet Exponents of Equivariant Orbits

Basic solutions of systems with an S^1-symmetry are rotating waves, which respect the symmetry, i.e., circular POs, which behave like

$$x = r\cos(\Omega t), \quad y = r\sin(\Omega t) \quad (r, \Omega \in \mathbb{R} \text{ const.})$$

for an appropriate choice of coordinates x and y and all other variables being constant. Such types of orbits are especially important in laser systems, because the laser equations are always invariant with respect to a phase shift of the complex electric field. In the laser systems these rotating wave solutions are then the modes of the laser. In case of a laser with feedback they are called external cavity modes (see Sect. 11.3).

For such orbits it is possible to find the transcendental equation for the Floquet exponents analytically. There are two essentially equivalent approaches to simplify the problem. The first method is to transform into a co-rotating frame

$$\hat{x} = \cos(\Omega t)x + \sin(\Omega t)y,$$
$$\hat{y} = -\sin(\Omega t)x + \cos(\Omega t)y.$$

In this frame the PO has become a circle of FPs. Since the PO respects the S^1-symmetry, each of the points has the same stability properties. The stability of the PO is then given by the stability of one of these FP. This method will be discussed in Sect. 13.2.

The other method uses the radius and phase as new coordinates. We use this approach to calculate the Floquet exponents of the Pyragas orbit. Rewriting (3.3) in polar coordinates $z(t) = r(t)\, e^{i\theta(t)}$ gives

$$\frac{d}{dt}r = (\lambda + r^2)r + b_0\left[\cos(\beta + \theta_\tau - \theta)r_\tau - r\cos\beta\right], \qquad (3.24a)$$

$$\frac{d}{dt}\theta = 1 + \gamma r^2 + b_0\left[\sin(\beta + \theta_\tau - \theta)\frac{r_\tau}{r} - \sin\beta\right]. \qquad (3.24b)$$

Linearizing around the PO according to $r(t) = r_0 + \delta r(t)$ and $\theta(t) = \Omega t + \delta\theta(t)$ with $r_0 = \sqrt{-\lambda}$ and $\Omega = 1 - \gamma\lambda$ we find

$$\frac{d}{dt}\begin{pmatrix}\delta r\\ \delta\theta\end{pmatrix} = \begin{bmatrix} -2\lambda - b_0\cos\beta & b_0 r_0\sin\beta \\ 2\gamma r_0 - b_0\sin\beta r_0^{-1} & -b_0\cos\beta \end{bmatrix}\begin{pmatrix}\delta r\\ \delta\theta\end{pmatrix}$$
$$+ \begin{bmatrix} b_0\cos\beta & -b_0 r_0\sin\beta \\ b_0\sin\beta r_0^{-1} & b_0\cos\beta \end{bmatrix}\begin{pmatrix}\delta r_\tau\\ \delta\theta_\tau\end{pmatrix}.$$

With the ansatz

$$\begin{pmatrix}\delta r(t)\\ \delta\theta(t)\end{pmatrix} = u\, \exp(\Lambda t),$$

where u is a two-dimensional constant vector, one obtains the autonomous linear equation

$$\begin{bmatrix} -2\lambda + b_0\cos\beta(e^{-\Lambda\tau} - 1) - \Lambda & -b_0 r_0\sin\beta(e^{-\Lambda\tau} - 1) \\ 2\gamma r_0 + b_0 r_0^{-1}\sin\beta(e^{-\Lambda\tau} - 1) & b_0\cos\beta(e^{-\Lambda\tau} - 1) - \Lambda \end{bmatrix} u = 0.$$

The vector u can only be mapped to 0 if the determinant of the matrix vanishes. This condition of vanishing determinant then gives the transcendental characteristic equation

$$0 = \left(-2\lambda + b_0\cos\beta(e^{-\Lambda\tau} - 1) - \Lambda\right)\left(b_0\cos\beta(e^{-\Lambda\tau} - 1) - \Lambda\right)$$
$$+ b_0 r_0\sin\beta(e^{-\Lambda\tau} - 1)\left(2\gamma r_0 + b_0 r_0^{-1}\sin\beta(e^{-\Lambda\tau} - 1)\right)$$

for the Floquet exponents Λ, which can be solved numerically.

Figure 3.9 depicts, for three different values of λ, the domain in the complex b-plane, where the Pyragas orbit is stable. The color shading shows the real part of the largest Floquet exponent and thus indicates the stability of the orbit. Large negative values correspond to enhanced stability. Outside the shaded area the PO is unstable. The black lines show the boundary of control in the limit $\lambda \nearrow 0$ according to (3.17). With increasing $|\lambda|$ the domain of control shrinks and for sufficiently large $|\lambda|$ stabilization is no longer possible. Note that for real valued b, i.e., $\beta = 0$, stabilization is not possible at all. Hence, a phase in the feedback gain is needed in order to stabilize the Pyragas orbit.

Fig. 3.9 Domain of control in the plane of complex feedback gain $b = b_0 e^{i\beta}$ for three different values of λ. The *black solid lines* indicate the boundary of stability in the limit $\lambda \nearrow 0$. The *color code* shows the largest (*negative*) real part of the periodic orbit's Floquet exponent. Parameters: $n = 1$, $\gamma = -10, \tau = 2\pi/(1 - \gamma\lambda)$

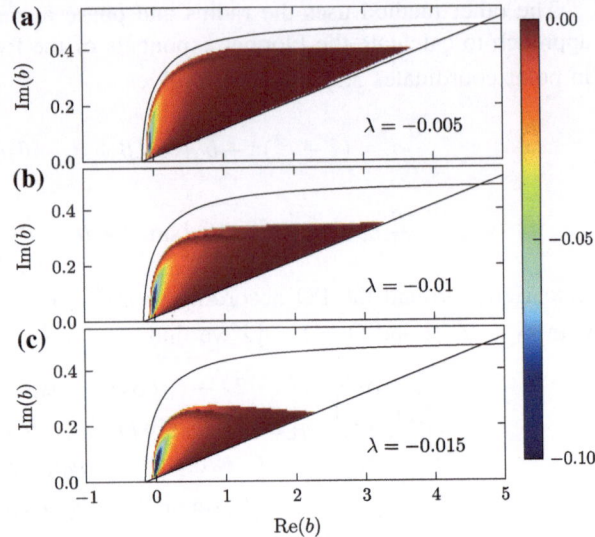

We will now investigate how the Pyragas orbit is stabilized in a more global picture for finite λ. In particular we will find delay induced orbits which bifurcate with the Pyragas orbit.

3.4 Stabilization Mechanism

Consider the situation, where we choose a feedback phase $\beta = \pi/4$ and turn up the feedback strength $b_0 = 0 \dots 0.3$, i.e., moving into the region of control. There has to be a mechanism, i.e., a bifurcation, which stabilizes the Pyragas orbit and transforms the subcritical Hopf bifurcation into a supercritical bifurcation as the feedback strength is increased.

To find this stabilization mechanism and obtain a comprehensive picture, we study all rotating wave solutions, which exist in the system. This is completely analogous to the standard method of calculating the external cavity modes of a laser system (see Sect. 11.3). We thus make the ansatz

$$r(t) = r_0, \qquad \theta(t) = \omega t$$

with constant r_0 and ω and insert this into ((3.24a), (3.24b))

$$0 = \lambda + r_0^2 + b_0 \Big[\cos(\beta + \omega\tau) - \cos\beta\Big],$$
$$\omega = 1 + \gamma r_0^2 + b_0 \Big[\sin(\beta + \omega\tau) - \sin\beta\Big].$$
$$\text{(3.25)}$$

Fig. 3.10 Radii r_0 (panel
(**a**)) and frequencies ω (panel
(**b**)) of rotating wave
solutions vs b_0. *Solid* and
dashed lines correspond to
stable and unstable solutions,
respectively. *Red lines*
indicate delay-induced orbits.
The *marked points* indicate
bifurcations of the solutions:
SN–saddle-node bifurcation,
TC–transcritical bifurcation,
subH–subcritical Hopf
bifurcation. Parameters:
$\lambda = -0.005, \gamma = -10,$
$\beta = \pi/4, \tau = 2\pi/(1 - \gamma\lambda)$

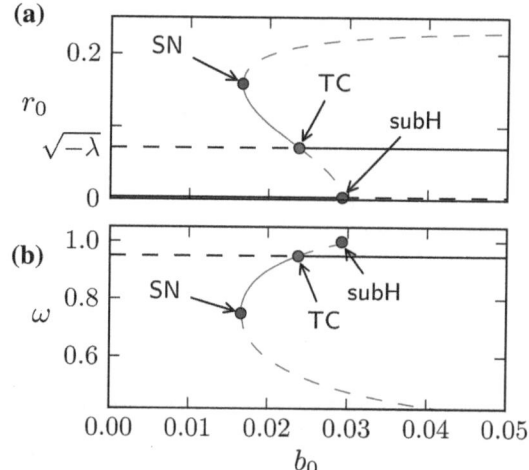

Eliminating r_0^2 gives a transcendental equation for the possible frequencies

$$\omega = 1 - \gamma\lambda + b_0\left[\sin(\beta + \omega\tau) - \sin\beta - \gamma\cos(\beta + \omega\tau) + \gamma\cos\beta\right].$$

This equation can be solved numerically and the obtained frequencies can then be inserted into either one of (3.25) to calculate r_0^2. Just as in the laser case, some obtained frequencies may result in negative r_0^2. These spurious solutions can be omitted. Figure 3.10 depicts the results of the calculations.

The radius $r_0 = \sqrt{-\lambda}$ and frequency $\omega = 1 - \gamma\lambda$ of the Pyragas orbit (black line) remain constant, since the control method is noninvasive on the target. With increasing control force b_0, however, two other delay-induced rotating wave solutions (red curves) are created in a saddle-node bifurcation (SN). At the transcritical bifurcation (TC) the Pyragas orbit and the stable delay-induced orbit exchange stability. The latter vanishes in a subcritical Hopf bifurcation (subH), at which the FP $z = 0$ becomes unstable. Note that there is a small interval of b_0 values, for which the Pyragas orbit and the fixed point are both stable. This is due to the finite value of λ. With decreasing $|\lambda|$ the transcritical and the Hopf bifurcation move closer together and in the limit of $\lambda \nearrow 0$ they coincide. Figure 3.11(a) depicts the dependence of the Pyragas orbit's Floquet exponents Λ and of the FP's eigenvalues η on the control amplitude b_0. With increasing b_0 the FP loses stability in the subcritical Hopf bifurcation (subH) shortly after the Pyragas orbit becomes stable in the transcritical bifurcation (TC) (compare Fig. 3.10). For larger values of b_0 a pair of complex-conjugate Floquet exponents (gray curve) crosses the imaginary axis ($\text{Re}(\Lambda\tau) = 0$) and destabilizes the Pyragas orbit in a torus bifurcation (T). The relevant Floquet multiplier $\mu = \exp(\Lambda\tau)$ at the torus bifurcation

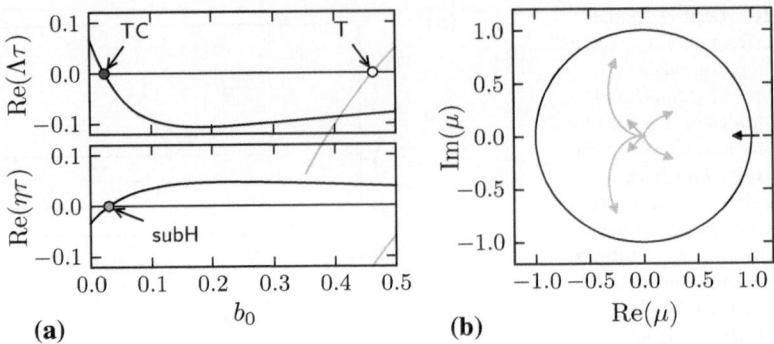

Fig. 3.11 Panel (**a**): Real parts of Floquet exponents Λ of the Pyragas orbit (*top*) and real part of the fixed points eigenvalue η (*bottom*) vs b_0. Panel (**b**): Floquet multipliers $\mu = \exp(\Lambda\tau)$ in the complex plane with the feedback amplitude $b_0 = [0, 0.3]$ as a parameter. The *arrows* indicate the direction of increasing b_0. Other parameters as in Fig. 3.10

are given by $\mu \approx e^{\pm i1.494\pi}$. Figure 3.11(b) depicts the Floquet multipliers in the complex plane with $b_0 = 0\ldots0.3$ as a curve parameter. With increasing b_0 the isolated odd-number multiplier with real part larger than one passes through the unit circle at 1. Note that there is also the Goldstone multiplier located at 1, which makes the crossing possible and had been overseen in [5]. Also, as b_0 is increased an infinite number of complex-conjugate multipliers indicated by the gray curves are generated by the control and move towards the unit circle, which then results in the torus bifurcation for larger values of b_0.

3.5 Conclusion and Discussion

In this section we have provided a counterexample, which refutes the odd-number theorem of time-delayed feedback control. In this example, of a subcritical Hopf bifurcation's normal form, we are able to stabilize the UPO, provided the system is sufficiently nonlinear, i.e., the absolute value of the parameter γ, which describes the dependence of the oscillation period on the amplitude, needs to be large enough. Furthermore, we saw that the control amplitude has to be complex. We will discuss this issue in more detail in Chaps. 5 and 6.

Recently, it was shown [6], using center manifold theory for DDEs and normal form analysis, that Pyragas control can also stabilize odd-number orbits in n-dimensional systems ($n > 2$) and that in fact for proper choices of the feedback matrix the system can be reduced to the normal form of (3.3). This justifies the approach we used above.

References

1. B. Fiedler, V. Flunkert, M. Georgi, P. Hövel, E. Schöll, Refuting the odd number limitation of time-delayed feedback control. Phys. Rev. Lett. **98**, 114101 (2007)
2. K. Engelborghs, T. Luzyanina, G. Samaey, DDE-BIFTOOL v.2.00: a matlab package for bifurcation analysis of delay differential equations, Tech. Rep. TW-330, Department of Computer Science, K.U. Leuven, Belgium (2001)
3. K. Engelborghs, T. Luzyanina, D. Roose, Numerical bifurcation analysis of delay differential equations using dde-biftool. ACM Trans. Math. Software. **28**, 1 (2002)
4. R. Szalai, Knut: A continuation and bifurcation software for delay-differential equations (2009)
5. H. Nakajima, On analytical properties of delayed feedback control of chaos. Phys. Lett. A. **232**, 207 (1997)
6. G. Brown, C.M. Postlethwaiten, M. Silber, Time-delayed feedback control of unstable periodic orbits near a subcritical hopf bifurcation Physica D. submitted (2010)

Chapter 4
Odd-Number Orbits Close to a Fold Bifurcation

The previous section showed that the alleged odd-number theorem is not valid for autonomous systems and that in fact an odd-number orbit born in a subcritical Hopf bifurcation can be stabilized by time-delayed feedback control. Although this is a very generic example, the question arises whether this situation close to the Hopf bifurcation is special or if odd-number orbits born from other bifurcations can be stabilized, too.

4.1 Model and Analysis

Following [1] we will now consider a PO born in a saddle-node (fold) bifurcation of POs. The normal form of this bifurcation is given by

$$\frac{d}{dt}z = \left[(|z|^2 - 1)^2 - \lambda + i\omega_0 + i\gamma(|z|^2 - 1) \right] z \qquad (z \in \mathbb{C}) \qquad (4.1)$$

with $\omega_0, \gamma > 0$. In polar coordinates $z(t) = r(t)\,e^{i\theta(t)}$ the normal form reads

$$\frac{d}{dt}r = [(r^2 - 1)^2 - \lambda]\,r =: g(\lambda, r)\,r, \qquad (4.2a)$$

$$\frac{d}{dt}\theta = \gamma(r^2 - 1) + \omega_0 =: h(\lambda, r). \qquad (4.2b)$$

The function g determines the radii of the rotating wave (RW) solutions and the function h determines the periods. At $\lambda = 0$ two RW solutions with $r = const$ and $\theta = \omega t$ are born in a saddle-node bifurcation at $r^2 = 1$. The corresponding bifurcation diagram is depicted in Fig. 4.1. From (4.2) we find the amplitudes $r_\pm$ and frequencies $\omega_\pm$ of these orbits

V. Flunkert, *Delay-Coupled Complex Systems*, Springer Theses,
DOI: 10.1007/978-3-642-20250-6_4, © Springer-Verlag Berlin Heidelberg 2011

Fig. 4.1 Bifurcation diagram
of the saddle-node bifurcation
of rotating waves. Solid and
dashed curves correspond to
stable and unstable solutions,
respectively

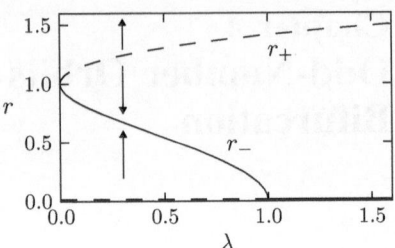

$$r_\pm = \sqrt{1 \pm \sqrt{\lambda}}, \qquad \omega_\pm = \omega_0 \pm \gamma\sqrt{\lambda}. \tag{4.3}$$

The sign "+" corresponds to the unstable (upper) branch and "−" corresponds
to the stable (lower) branch. At $\lambda = 1$ the stable branch vanishes in a supercritical
Hopf bifurcation stabilizing the FP $z = 0$.

We will now stabilize the "+" branch, which is obviously an odd-number orbit,
by applying time-delayed feedback to the system according to

$$\frac{d}{dt} z = \big[g(\lambda, |z|) + i\, h(\lambda, |z|)\big]\, z + b\big[z_\tau - z\big]. \tag{4.4}$$

Again, for noninvasive control the delay time has to be chosen as

$$\tau = k\, T_+ = k\, \frac{2\pi}{\omega_+} = \frac{2\pi k}{\omega_0 + \gamma\sqrt{\lambda}} \tag{4.5}$$

to match the period of the "+" branch. Solving for λ we obtain the Pyragas curve
in the (τ, λ)-plane

$$\lambda_P(\tau) = \left(\frac{2\pi k - \tau\omega_0}{\gamma\tau}\right)^2. \tag{4.6}$$

Figure 4.2 depicts the Pyragas curves for different integer numbers k.

We now consider τ as the relevant bifurcation parameter and adjust λ according
to (4.6). In the uncontrolled case $b = 0$, we obtain the RWs

$$r = \sqrt{1 \pm \frac{2\pi k - \omega_0\tau}{\gamma\tau}}, \qquad \omega = \omega_0 \pm \frac{2\pi k - \omega_0\,\tau}{\tau}. \tag{4.7}$$

The bifurcation diagram of these branches is depicted in Fig. 4.3. At $\tau = T_0 = 2\pi/\omega_0$ the two branches form a transcritical bifurcation (TC). This transcritical
bifurcation at first seems artificially introduced, by moving in Fig. 4.1 according to
$\lambda_P(\tau)$ down the λ-axis to $\lambda = 0$ and then up again. However, for $b_0 \neq 0$ the blue
branch in Fig. 4.3 is unchanged because it features $\tau = T$, while the other branch is
affected by the control and thus changes. If it is now possible to shift this branch

Fig. 4.2 (*Color online*) Pyragas curves in the (τ, λ)-plane for different values $k = 1, 2, \ldots, 11$. Each Pyragas curve touches the dashed (*blue*) saddle-node line (SN) $\lambda = 0$ at a transcritical bifurcation point (TC). Parameters: $\gamma = 1$, $\omega_0 = 1$

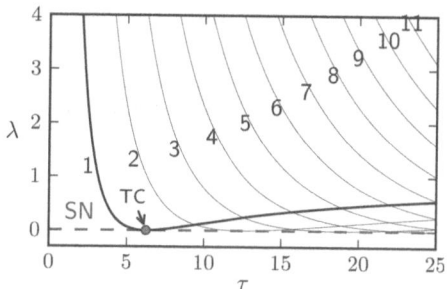

Fig. 4.3 Transcritical bifurcation for $\lambda = \lambda_P(\tau)$. The gray line is the Pyragas branch. *Solid* and *dashed* lines correspond to stable and unstable solutions, respectively. Parameters: $\gamma = 10$, $\omega_0 = 1$, $T_0 = 2\pi/\omega_0 = 2\pi$

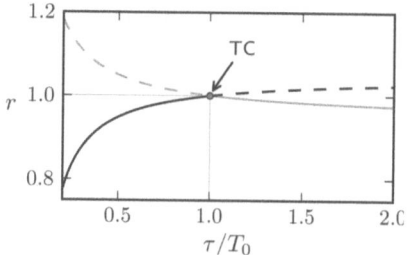

upwards and thus the transcritical bifurcation $\tau = \tau_c$ to the left, then the Pyragas branch is stable also for $r > 0$, i.e., the upper branch of the saddle-node bifurcation has been stabilized.

To approach this problem, we rewrite the controlled system (4.4) in polar coordinates

$$\frac{d}{dt}r = \left[(r^2 - 1)^2 - \lambda\right]r + b_0[\cos(\beta + \theta_\tau - \theta)\, r_\tau - \cos\beta\, r],$$
$$\frac{d}{dt}\theta = \gamma(r^2 - 1) + \omega_0 + b_0[\sin(\beta + \theta_\tau - \theta)\, r_\tau/r - \sin\beta].$$

(4.8a)

The RW solutions then obey

$$0 = \varepsilon^2 - \lambda + 2b_0 \sin(\omega\tau/2)\sin(\beta - \omega\tau/2) \tag{4.9a}$$

$$\omega = \gamma\varepsilon + \omega_0 - 2b_0 \sin(\omega\tau/2)\cos(\beta - \omega\tau/2), \tag{4.9b}$$

with $\varepsilon := r^2 - 1$. Solving (4.9b) for ε

$$\varepsilon = \gamma^{-1}\left[\omega - \omega_0 + 2b_0 \sin(\omega\tau/2)\cos(\beta - \omega\tau)\right]$$

we see that the right hand side increases monotonically in ω for small b_0, since in this case the oscillating part is small compared to the linearly increasing ω. Then there exists an inverse function $\omega(\varepsilon)$. Inserting this into (4.9a) yields

$$0 = G(\tau, \varepsilon) := \varepsilon^2 - \lambda + 2b_0 \sin(\omega(\varepsilon)\tau/2)\sin(\beta - \omega(\varepsilon)\tau/2).$$

At the transcritcal bifurcation $\tau = \tau_c$ the radius equation (4.9a) has a two-fold root, i.e.,

$$0 = \partial_\varepsilon G(\tau_c, \varepsilon) \quad \text{and} \quad \partial_\varepsilon^2 G(\tau_c, \varepsilon) \neq 0,$$

in addition to $G(\tau_c, \varepsilon) = 0$. Evaluating this equation on the Pyragas branch, where $\omega\tau = 2\pi\tau/T = 2\pi k$, we obtain

$$\begin{aligned} 0 = \partial_\varepsilon G(\tau_c, \varepsilon) &= 2\varepsilon + b_0\, \tau_c \cos(k\pi) \sin(\beta - k\pi)\, (\partial_\varepsilon\omega) \\ &= 2\varepsilon + b_0\, \tau_c(\partial_\varepsilon\omega) \sin\beta. \end{aligned} \quad (4.10)$$

To obtain the unknown function $\partial_\varepsilon\omega$, we implicitly differentiate (4.9b) with respect to ε at $\omega\tau = 2\pi k$

$$\partial_\varepsilon\omega = \gamma - b_0\, \tau(\partial_\varepsilon\omega) \cos\beta$$

and solve for $\partial_\varepsilon\omega$

$$\partial_\varepsilon\omega = \frac{\gamma}{1 + b_0\, \tau \cos\beta} = \frac{\gamma}{1 + b_0 \frac{2\pi k}{\omega_0 + \gamma\varepsilon} \cos\beta}.$$

Here we have used $\omega\tau = 2\pi k$ and $\omega = \omega_0 + \gamma\varepsilon$. Inserting this into (4.10) yields

$$0 = \varepsilon(\omega_0 + \gamma\varepsilon + b_0 2\pi k \cos\beta) + b_0 \pi k \gamma \sin\beta$$

Solving for b_0 then gives the control force b_0 at the bifurcation

$$b_c = -\varepsilon \frac{\omega_0 \gamma \varepsilon}{\pi k(\gamma \sin\beta + 2\varepsilon \cos\beta)}. \quad (4.11)$$

An equivalent condition involving τ_c can be found by substitution of (4.5) and $-\sqrt{\lambda} = r^2 - 1 = \varepsilon$

$$b_c = -\frac{1}{\tau_c} \cdot \frac{2\pi k - \omega_0\tau_c}{\frac{1}{2}\gamma^2\tau_c \sin\beta + (2\pi k - \omega_0\tau_c) \cos\beta}. \quad (4.12)$$

From (4.11) and (4.12), for small ε ($\tau_c \approx 2\pi k/\omega_0$), it follows that the optimal control phase, i.e., the phase with smallest $|b_c|$, is $\beta = -\pi/2$. For this optimal control phase (4.11) and (4.12) are simplified to

$$b_c = \frac{\varepsilon}{\pi k}\left(\frac{\omega_0}{\gamma} + \varepsilon\right) \quad (4.13)$$

and

$$b_c = \frac{2}{(\gamma\tau_c)^2}\left(2\pi k - \omega_0\tau_c\right), \quad (4.14)$$

respectively. Solving (4.13) and (4.14) for ε and τ_c, respectively, and expanding for small $b_0 > 0$, we find the location of the transcritical bifurcation

$$\varepsilon = -\left(\pi k \frac{\gamma}{\omega_0} \sin \beta\right) b_0 + \mathcal{O}(b_0^2),$$

$$\tau_c = \frac{2\pi k}{\omega_0} + \left[\frac{1}{2\omega_0}\left(\frac{2\pi k\gamma}{\omega_0}\right)^2 \sin \beta\right] b_0 + \mathcal{O}(b_0^2).$$

From this analysis we can conclude that stabilization of the unstable odd-number orbit is possible. Near the fold for $\gamma > 0$ and $\sin \beta < 0$ stabilization can be realized by arbitrarily small control amplitudes b_0.

4.2 Stabilization Mechanism

Similar to the analysis in Sect. 3.4, we can obtain a comprehensive picture of the bifurcation scenario by looking at all RW present in the system. The RW ansatz results in (see (4.9))

$$0 = (r^2 - 1)^2 - \lambda + 2b_0 \sin(\omega\tau/2) \sin(\beta - \omega\tau/2),$$
$$\omega = \gamma(r^2 - 1) + \omega_0 - 2b_0 \sin(\omega\tau/2) \cos(\beta - \omega\tau/2).$$

Eliminating r^2 gives a transcendental equation for the frequencies

$$0 = -\gamma^2\lambda + \gamma^2 2b_0 \sin(\omega\tau/2) \sin(\beta - \omega\tau/2)$$
$$+ \left[\omega - \omega_0 + 2b_0 \sin(\omega\tau/2) \cos(\beta - \omega\tau/2)\right]^2.$$

Solving this equation numerically and inserting the frequencies into

$$r = \sqrt{\frac{\omega - \omega_0}{\gamma} - \frac{2b_0}{\gamma} \sin\left(\frac{\omega\tau}{2}\right) \cos\left(\beta - \frac{\omega\tau}{2}\right) + 1}$$

we obtain the allowed radii, after eliminating spurious (imaginary) solutions. The bifurcation diagram obtained from these RW solutions is depicted in Fig. 4.4. The orbit, which stabilizes the Pyragas orbit, may be the minus branch or another delay-induced orbit, which is born from a fold bifurcation, depending on the parameters. The crossover between these two scenarios occurs at $\gamma \approx 10.6$, where two saddle-node points merge and disappear in a transcritical bifurcation. The radius of the Pyragas orbit does not change with b_0, since the control is noninvasive on this orbit. The "$-$"-branch on the other hand is affected and the radius changes.

Fig. 4.4 Radii of stable (*solid*) and unstable (*dotted*) rotating solutions vs b_0 for different values of γ. *Squares* and *circles* indicate saddle-node and transcritical bifurcations, respectively. Parameters: $\tau = 2\pi/(1 - \gamma\lambda)$, $\lambda = 0.001$, $\omega_0 = 1$, $\beta = -\pi/2$

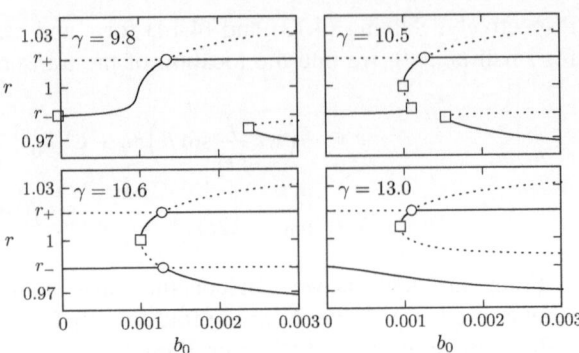

4.3 Domain of Control

To calculate the domain of control numerically we proceed in the same way as in Sect. 3.3. We linearize (4.8) around the Pyragas orbit according to $z(t) = (r + \delta r(t)) \exp\left(i\omega t + i\delta\varphi(t)\right)$

$$\frac{d}{dt}\begin{pmatrix} \delta r \\ \delta\varphi \end{pmatrix} = \begin{bmatrix} \partial_r g\, r + g - b_0 \cos\beta & rb_0 \sin(\beta - \omega\tau) \\ \partial_r h - b_0 \sin(\beta - \omega\tau)/r & -b_0 \cos(\beta - \omega\tau) \end{bmatrix}\begin{pmatrix} \delta r \\ \delta\varphi \end{pmatrix}$$
$$+ \begin{bmatrix} b_0 \cos(\beta - \omega\tau) & -rb_0 \sin(\beta - \omega\tau) \\ b_0 \sin(\beta - \omega\tau)/r & b_0 \cos(\beta - \omega\tau) \end{bmatrix}\begin{pmatrix} \delta r_\tau \\ \delta\varphi_\tau \end{pmatrix}.$$

The delay time τ matches the period of the Pyragas orbit and we thus have

$$\omega\tau = 2\pi k.$$

Using the exponential ansatz $(\delta r(t), \delta\varphi(t)) \propto \exp(\Lambda t)$ yields a transcendental characteristic equation $0 = \chi(\Lambda)$ for the Floquet exponents Λ

$$0 = \chi(\Lambda) = \det\begin{bmatrix} M_{11} & M_{12} \\ M_{21} & M_{22} \end{bmatrix} = 0 \tag{4.15}$$

with

$$M_{11} = 4(r_+^2 - 1)r_+^2 + (r_+^2 - 1)^2 - \lambda - \Lambda - \left(1 - e^{-\Lambda\tau}\right)b_0 \cos\beta,$$
$$M_{12} = r_+\left(1 - e^{-\Lambda\tau}\right)b_0 \sin\beta,$$
$$M_{21} = 2\gamma r_+ - \left(1 - e^{-\Lambda\tau}\right)b_0 \sin(\beta)/r_+,$$
$$M_{22} = -\Lambda - \left(1 - e^{-\Lambda\tau}\right)b_0 \cos\beta.$$

Solving this equation numerically, we obtain the control domain in the complex b-plane. Figure 4.5 depicts this domain in polar coordinates $b = b_0\, e^{i\beta}$ (panel (a)) and Cartesian coordinates $b = b_r + i\, b_i$ (panel (b)). The color code shows the largest (negative) real part of the Floquet exponent. The control domain is bounded on one side by the transcritical bifurcation line and on the other side by a line of

Fig. 4.5 Domain of control
(a) in the (β, b_0)-plane and
(b) in the (b_r, b_i)-plane. The
color code shows only
negative values of the largest
real part of the Floquet
exponents. Parameters:
$\omega_0 = 1.0$, $\lambda = 10^{-4}$, $\gamma = 0.1$,
$\tau = 2\pi/(1 - \gamma/\lambda)$

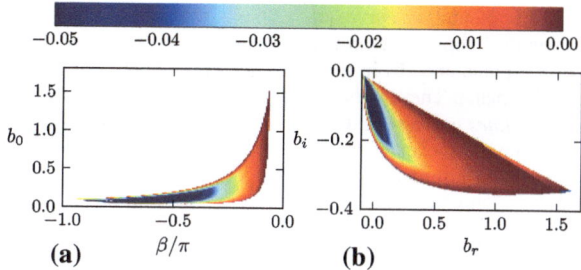

secondary Hopf bifurcations, i.e., torus bifurcations, which destabilize the Pyragas
orbit for large values of the control amplitude.

The transcritical bifurcation line can be calculated analytically from (4.15) as
follows. A transcritical bifurcation occurs if real a Floquet exponent changes sign,
i.e., $\Lambda = 0$. Since we have an autonomous system, one Floquet exponent is always
zero corresponding to the Goldstone mode of the orbit. To obtain the transcritical
bifurcation line, we will use a series expansion of the characteristic equation in
$x = -\Lambda\tau$ up to order x^2. At the transcritical bifurcation the solution $x = 0$ then has
multiplicity two.

The following calculations are correct up to second order in x. Using

$$1 - e^{-\Lambda\tau} = -x - \frac{1}{2}x^2, \qquad \Lambda = -\frac{1}{\tau}x,$$

we obtain

$$0 = \det \begin{bmatrix} a + \frac{1}{\tau}x + \left(x + \frac{1}{2}x^2\right)b_0 \cos\beta & -r\left(x + \frac{1}{2}x^2\right)b_0 \sin\beta \\ c + \frac{1}{r}\left(x + x^2\right)b_0 \sin\beta & \frac{1}{\tau}x + \left(x + x^2\right)b_0 \cos\beta \end{bmatrix},$$

where we used the abbreviations

$$a = 4(r^2 - 1)r^2 + (r^2 - 1)^2 - \lambda, \qquad c = 2\gamma r.$$

Calculating this determinant (up to second order) gives

$$0 = a\frac{1}{\tau}x + a(x + x^2)\,b_0 \cos\beta + \frac{1}{\tau^2}x^2 + \frac{2}{\tau}x^2\,b_0 \cos\beta + x^2\,b_0^2 \cos^2\beta$$
$$+ cr(x + x^2)\,b_0 \sin\beta + x^2\,b_0^2 \sin^2\beta.$$

Factorizing out the Goldstone mode $x = 0$ yields

$$0 = \frac{1}{\tau}a + ab_0 \cos\beta + crb_0 \sin\beta + \left[\frac{1}{\tau^2} + b_0^2 + \left(ab_0 + \frac{2}{\tau}b_0\right)\cos\beta + crb_0 \sin\beta\right]x,$$

which then gives the transcritical bifurcation for $x = 0$

Fig. 4.6 Boundary curves of
the control domain (**a**) in the
(β, b_0)-plane and (**b**) in the
(b_r, b_i)-plane. The *solid* and
dashed lines correspond to
the Neimark-Sacker and
transcritical bifurcation line.
Parameters as in Fig. 4.5

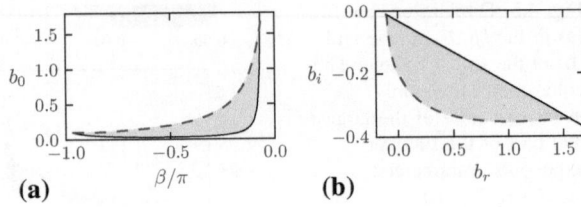

$$b_0 = -\frac{a}{\tau} \cdot \frac{1}{a\cos\beta + cr\sin\beta}. \tag{4.16}$$

The other side of the control domain is bounded by a Hopf bifurcation line, where the Pyragas orbit loses its stability in a Neimark-Sacker bifurcation. Since it is not possible to find this line analytically, we follow this branch of the solution numerically. The two resulting curves are shown in Fig. 4.6.

4.4 Conclusion

We have shown that odd-number orbits can also be stabilized close to a saddle-node bifurcation of POs. The stabilization mechanism is again a transcritical bifurcation with a delay-induced stable orbit. Similarly as in Chap. 3, stabilization is possible close to the bifurcation, if the system is sufficiently nonlinear and if the feedback amplitude is complex.

In this and the previous section, we have shown that the odd-number limitation is not valid for the two situations of a subcritical Hopf bifurcation and a saddle-node bifurcation of PO. This result mathematically refutes the alleged odd-number theorem. However, these counter examples have a very particular feedback term, which preserves the rotation symmetry of the uncontrolled equations. In the next section we will look at other feedback terms which break the symmetry, but are more readily applicable for experimental realization.

References

1. B. Fiedler, S. Yanchuk, V. Flunkert, P. Hövel, H.J. Wünsche, E. Schöll, Delay stabilization of rotating waves near fold bifurcation and application to all-optical control of a semiconductor laser. Phys. Rev. E **77**, 066207 (2008)

Chapter 5
Towards Stabilization of Odd-Number Orbits in Experiments

Although the counterexample provided in Chap. 3 has fulfilled its purpose to refute the alleged odd-number limitation, it is often difficult to realize in experiments, in order to stabilize odd-number orbits, and in particular subcritical Hopf orbits.

One reason why the counterexample is not immediately applicable, is the special choice of the gain matrix, i.e., a feedback term, which only involves z and not the complex conjugate $\bar{z}$. This gain matrix conserves the S^1-symmetry of the normal form, but in order to realize this control matrix experimentally one needs to have access to two dynamical variables in the rotation plane of the orbit, process these to generate the rotation phase β, and feed the control signal back into the corresponding two dynamic degrees of freedom. This may be possible in certain situations, for instance, when stabilizing an unstable mode of a laser, where the optical phase can naturally introduce a rotation [1] (see Sect. 7.1). But what happens, for example, if we have only access to one dynamical variable? We will give some answers to this question in the following [2].

5.1 Model

Consider a dynamical system with a bifurcation parameter μ, which undergoes a subcritical Hopf bifurcation at $\mu = 0$ with the UPO lying without loss of generality on the $\mu < 0$ side. The center manifold theorem implies that close to the bifurcation the system equations can be transformed to the normal form

$$\frac{d}{dt}\begin{pmatrix} x \\ y \end{pmatrix} = \begin{bmatrix} d\mu + ar^2 & -(\omega + c\mu + br^2) \\ \omega + c\mu + br^2 & d\mu + ar^2 \end{bmatrix}\begin{pmatrix} x \\ y \end{pmatrix}, \tag{5.1}$$

with $r^2 = x^2 + y^2$ or in complex notation $(z = x + iy)$

$$\frac{d}{dt}z = \left[(d + ic)\mu + i\omega + (a + ib)|z|^2 \right]z.$$

V. Flunkert, *Delay-Coupled Complex Systems*, Springer Theses,
DOI: 10.1007/978-3-642-20250-6_5, © Springer-Verlag Berlin Heidelberg 2011

We choose $d > 0$ and $a > 0$. This means the FP is stable or unstable for $\mu < 0$ or $\mu > 0$, respectively, and the UPO lies on the $\mu < 0$ side. For simplicity we will assume $\omega > 0$. The case $\omega < 0$ is easily recovered by exchanging the variables $x \longleftrightarrow y$.

Although these equations can be simplified further to the form of (3.1)

$$\frac{d}{dt} z = \left[\lambda + i + (1 + i\gamma)|z|^2 \right] z$$

by rescaling of the bifurcation parameter, the time, and the dynamical variables, we keep this form of the equations to allow easier comparison with experimental situations. In particular we calculate the normal form coefficients ω, a, b, c, and d for a laser model in Sect. 7.2.

In polar coordinates the equations are given by

$$\frac{d}{dt} r = (d\mu + a r^2)r,$$

$$\frac{d}{dt} \theta = (\omega + c\mu + br^2),$$

and the radius and period of the UPO can be read off

$$r = \sqrt{-\frac{d}{a}\mu},$$

$$T = \frac{2\pi}{|\omega + c\mu + br^2|} = \frac{2\pi}{\left|\omega + \left(c - \frac{bd}{a}\right)\mu\right|}.$$

Let us now consider Pyragas feedback with a general coupling matrix K

$$\frac{d}{dt}\begin{pmatrix} x \\ y \end{pmatrix} = \begin{bmatrix} d\mu + a r^2 & -(\omega + c\mu + b r^2) \\ \omega + c\mu + b r^2 & d\mu + a r^2 \end{bmatrix}\begin{pmatrix} x \\ y \end{pmatrix} + \begin{bmatrix} K_{11} & K_{12} \\ K_{21} & K_{22} \end{bmatrix}\begin{pmatrix} x_\tau - x \\ y_\tau - y \end{pmatrix}.$$

$$(5.2)$$

We follow the proof idea of the counterexample (Chap. 3) and analyze the stability of the FP. Making the ansatz $(x, y) = u\, e^{\eta t}$, where u is a constant vector, we obtain the transcendental characteristic equation $\chi(\eta) = 0$ with

$$\chi(\eta) := \det\begin{bmatrix} d\mu - \eta + K_{11}F(\eta) & -\omega - c\mu + K_{12}F(\eta) \\ \omega + c\mu + K_{21}F(\eta) & d\mu - \eta + K_{22}F(\eta) \end{bmatrix},$$

where $F(\eta) = e^{-\eta\tau} - 1$. Calculating the determinant yields

$$\chi(\eta) = (d\mu - \eta)^2 + \operatorname{tr} K (d\mu - \eta)F(\eta) + \det K\, F(\eta)^2$$
$$+ (\omega + c\mu)^2 + \kappa(\omega + c\mu)F(\eta).$$

$$(5.3)$$

Here, we have introduced the parameter $\kappa := K_{21} - K_{12}$ that is a measure for the antisymmetry of the feedback matrix and will play a crucial role in the following analysis. Note that when we recover the case of negative ω by exchanging x and y as discussed above, κ changes sign in the characteristic equation $\kappa \rightarrow -\kappa$.

To follow the same argument as in [3] and Chap. 3, we need three ingredients:

1 The location of the Hopf points on the τ-axis (Hopf A and B points),
2 the crossing direction of the Hopf eigenvalues at these Hopf points, when going up the τ-axis,
3 the slope of the Hopf curve and the Pyragas curve at the Hopf A points.

With theses three ingredients we can construct parameters (feedback gain) such that there is a change from a (0)-region to a (2)-region along the Pyragas curve. The number in parentheses denotes again the total number of eigenvalues of the FP with $\text{Re}(\eta) > 0$.

Let us start by trying to find the Hopf points on the τ-axis. At $\mu = 0$, $\eta = i\Omega$ the real and imaginary part of (5.3) are

$$
\begin{aligned}
0 = \det K - \Omega^2 - \kappa\omega + \omega^2 + (\kappa\omega - 2 \det K) \cos(\Omega\tau) \\
+ \det K \cos(2\Omega\tau) - \Omega \text{tr} K \sin(\Omega\tau),
\end{aligned}
\tag{5.4a}
$$

$$
\begin{aligned}
0 = \Omega \text{tr} K (1 - \cos(\Omega\tau)) - (\kappa\omega - 2 \det K) \sin(\Omega\tau) \\
- \det K \sin(2\Omega\tau).
\end{aligned}
\tag{5.4b}
$$

In general (5.4) cannot be solved exactly. So we will restrict the analysis to a special class of feedback matrices.

5.2 Experimentally Relevant Feedback Matrices

Consider the following experimental situation. We are able to measure an output variable u of the system and are able to apply the control to an input variable v. After the center manifold reduction and normal form transformation u and v are functions of x and y, which we expand to the leading linear order

$$
\begin{aligned}
u = u(x, y) = u_1 x + u_2 y + \cdots, \\
v = v(x, y) = v_1 x + v_2 y + \cdots.
\end{aligned}
$$

Here, we have omitted constant terms, since they would disappear in the Pyragas feedback. We can picture the vectors (u_1, u_2) and (v_1, v_2) as being tangent to the center manifold at $(x, y) = (0, 0)$. The measured signal m is then given by the projection

$$
m(t) = \begin{pmatrix} u_1 \\ u_2 \end{pmatrix} \cdot \begin{pmatrix} x(t) \\ y(t) \end{pmatrix}
$$

and our control signal acts as

$$\frac{d}{dt}x = \cdots + v_1[m_\tau - m],$$

$$\frac{d}{dt}y = \cdots + v_2[m_\tau - m]$$

on the dynamical equations. This leads to the following gain matrix

$$K = \begin{bmatrix} v_1 u_1 & v_1 u_2 \\ v_2 u_1 & v_2 u_2 \end{bmatrix},$$

which has vanishing determinant. With $\det K = 0$ the characteristic (5.3) simplifies to

$$0 = (d\mu - \eta)^2 + \operatorname{tr} K(d\mu - \eta)F(\eta) + (\omega + c\mu)^2 + \kappa(\omega + c\mu)F(\eta). \qquad (5.5)$$

For this simpler equation it is now possible to carry out the analysis.

5.3 Analysis

(1) *Location of Hopf points*—To find the location of the Hopf points on the τ-axis, we insert $\eta = i\Omega$ into (5.5), set $\mu = 0$ and split the equation into real and imaginary parts

$$0 = -\Omega^2 - \kappa\omega + \omega^2 + \kappa\omega\cos(\Omega\tau) - \Omega\operatorname{tr} K\sin(\Omega\tau),$$
$$0 = \Omega\operatorname{tr} K(1 - \cos(\Omega\tau)) - \kappa\omega\sin(\Omega\tau). \qquad (5.6)$$

In the following we will for simplicity consider Ω to be positive. The complex conjugate solution is simply $\eta = -i\Omega$.
Writing (5.6) as

$$\begin{pmatrix} \Omega^2 - \omega^2 + \kappa\omega \\ -\Omega\operatorname{tr} K \end{pmatrix} = \begin{bmatrix} \cos\xi & \sin\xi \\ -\sin\xi & \cos\xi \end{bmatrix} \begin{pmatrix} \kappa\omega \\ -\Omega\operatorname{tr} K \end{pmatrix},$$

with $\xi = \Omega\tau$, it is obvious that there can only be a solution if the two vectors have the same length

$$(\Omega^2 - \omega^2 + \kappa\omega)^2 + \Omega^2\operatorname{tr} K^2 = \kappa^2\omega^2 + \Omega^2\operatorname{tr} K^2,$$

since the rotation matrix leaves the length of vectors invariant. This gives the values for Ω^2 on the τ-axis

$$\Omega^2 = \omega^2, \qquad \Omega^2 = \omega^2 - 2\kappa\omega.$$

With these Ω-values we can calculate the rotation angle ξ and the delay time τ. In particular for $\Omega^2 = \omega^2$ we recover the Pyragas points (alias A series) on the τ-axis

$$\tau_n^A = \frac{2\pi n}{\omega}, \qquad \Omega^A = \omega.$$

Inserting $\Omega = \Omega^B := \sqrt{\omega^2 - 2\kappa\omega}$ gives the B series. Note that $\kappa < \omega/2$ is necessary in order for Hopf B points to exist. When calculating ξ (and τ) for the Hopf B points we have to take into account the different possible signs of κ and $\operatorname{tr} K$:

$$\tau_n^B = \begin{cases} \frac{1}{\Omega^B}\left[2\pi n + \varphi\right], & \text{if } \kappa \operatorname{tr} K \geq 0, \\ \frac{1}{\Omega^B}\left[2\pi n + (2\pi - \varphi)\right], & \text{if } \kappa \operatorname{tr} K < 0, \end{cases} \tag{5.7}$$

with

$$\varphi = \arccos\left[\frac{(\operatorname{tr} K)^2(\omega^2 - 2\kappa\omega) - \omega^2\kappa^2}{(\operatorname{tr} K)^2(\omega^2 - 2\kappa\omega) + \omega^2\kappa^2}\right]. \tag{5.8}$$

The index is chosen such that $n = 0$ labels the first point in the series, i.e., $n = 0$ is the lowest integer with $\xi_n^B > 0$. The Hopf B series is spread equidistantly on the τ-axis with a distance

$$\Delta\tau^B = \tau_{n+1}^B - \tau_n^B = \frac{2\pi}{\sqrt{\omega^2 - 2\kappa\omega}}.$$

   (2) *Crossing direction of Hopf eigenvalue pair*—The crossing direction of the Hopf eigenvalues for Hopf points on the τ-axis is given by

$$\operatorname{sgn} \operatorname{Re}(\partial_\tau \eta) = \operatorname{sgn}\Big[-\Omega^2(\operatorname{tr} K)^2 + \Omega^2\big((\operatorname{tr} K)^2 + 2\kappa\omega\big)\cos(\Omega\tau)$$
$$+ \Omega \operatorname{tr} K\big(\kappa\omega - 2\Omega^2 \operatorname{tr} K\big)\sin(\Omega\tau)\Big]. \tag{5.9}$$

At the Pyragas points ($\Omega = \Omega^A$, $\tau = \tau_n^A$) this gives

$$\operatorname{sgn} \operatorname{Re}(\partial_\tau \eta)\Big|_A = \operatorname{sgn} \kappa.$$

The crossing direction of the Hopf eigenvalues at the Hopf B points ($\Omega = \Omega^B$, $\tau = \tau_n^B$) on the other hand is given by

$$\operatorname{sgn} \operatorname{Re}(\partial_\tau \eta)\Big|_B = \operatorname{sgn}\Big[2(\operatorname{tr} K)^2\kappa^2 - (\operatorname{tr} K)^2\kappa\omega - \kappa\omega^3\Big]$$
$$= -\operatorname{sgn} \kappa \operatorname{sgn}\Big[\omega^3 + (\operatorname{tr} K)^2(\omega - 2\kappa)\Big].$$

For the allowed κ-values ($\kappa < \omega/2$) this expression reduces to

$$\mathrm{sgn}\ \mathrm{Re}(\partial_\tau \eta) = -\mathrm{sgn}\kappa,$$

and hence the crossing direction is opposite to that at the Pyragas points.

(3) *Slope of Hopf and Pyragas curve*—By implicit differentiation of the characteristic (5.5) with respect to μ we find the slope of the Hopf curve at the Pyragas points

$$\partial_\mu \tau_H \Big|_{\tau=\tau_n^A} = -2\frac{d(n\pi\ \mathrm{tr}\,K + \omega) + cn\pi\kappa}{\omega^2\kappa} \tag{5.10}$$

The slope of the Pyragas curve at $\mu = 0$ is given by

$$\partial_\mu \tau_P = -\frac{2\pi n}{\omega^2}(c - bd/a)$$

Putting the pieces together we have to carefully distinguish different cases of different sign combinations of the various parameters. In particular we have to distinguish between the case

$$-(c - bd/a) < 0,$$

where the period of the UPO increases with increasing distance from the bifurcation (increasing period case) and the case

$$-(c - bd/a) > 0,$$

where it decreases with increasing distance from the bifurcation. Note that this distinction is not exactly the same as soft-/ and hard spring case (see p. xx), because the parameter c, which changes the period with the bifurcation parameter, can overrule the other terms bd/a, which changes the period with the amplitude of the oscillations.

For $-(c - bd/a) < 0$ the period of the orbit increases with increasing distance from the bifurcation and the Pyragas curve emanates to the upper left from the Pyragas point. For stabilization we need a (2)-region above and a (0)-region below the n-th Pyragas point. This means the eigenvalue crossing direction has to be positive at the A points and negative at the B points, i.e., $\kappa > 0$.

The Pyragas curve has to lie above the Hopf curve for $\mu < 0$, which means that the slopes at $\mu = 0$ have to obey $\partial_\mu \tau_p < \partial_\mu \tau_H$. This gives

$$\mathrm{tr}\,K < -\frac{b}{a}\kappa - \frac{\omega}{\pi n}. \tag{5.11}$$

Finally the order of the Hopf points has to be $\tau_{n-1}^B \le \tau_n^A$. Inserting the calculated τ-values (see (5.7)) gives two cases.

1. For $\mathrm{tr}\,K \ge 0$ inserting the τ-values gives

$$\varphi \le 2\pi + 2\pi n\left(\frac{\Omega^B}{\omega} - 1\right). \tag{5.12}$$

Depending on the values of n, κ and ω the right hand side may be negative and the inequality cannot be fulfilled since $\varphi \in [0, \pi]$. Stabilization is only possible if the right hand side is positive, i.e., if

$$\kappa \leq \omega \frac{2n - 1}{2n^2}.$$

Note that for $n = 1$ this coincides with our initial condition $\kappa < \omega/2$. For this valid κ-range the inequality (5.12) gives when inserting φ from (5.8) a condition on the magnitude of $\operatorname{tr} K$

$$\operatorname{tr} K \geq \frac{\kappa \omega}{\Omega^B} \cot\left(\pi n \sqrt{1 - 2\kappa/\omega}\right).$$

2. For $\operatorname{tr} K < 0$ we find

$$\varphi \geq 2\pi n \left(1 - \frac{\Omega^B}{\omega}\right). \tag{5.13}$$

Inserting φ then gives the same bound as above

$$\operatorname{tr} K \geq \frac{\kappa \omega}{\Omega^B} \cot\left(\pi n \sqrt{1 - 2\kappa/\omega}\right).$$

However, in this case the κ-domain is different and the left hand side as well as the right hand side are negative.

The necessary condition for the Hopf point ordering can, for the increasing period case with $\kappa > 0$ thus be summarized by

$$\kappa \leq \omega \frac{2n - 1}{2n^2},$$

$$\operatorname{tr} K \geq \frac{\kappa \omega}{\sqrt{\omega^2 - 2\kappa\omega}} \cot\left(\pi n \sqrt{1 - \frac{2\kappa}{\omega}}\right).$$

For $-(c - bd/a) > 0$ the period of the orbit decreases with increasing distance from the bifurcation and the Pyragas curve emanates to the lower left from the Pyragas point. For stabilization we need a (0)-region above and a (2)-region below the emanating Pyragas point. This means the eigenvalue crossing directions has to be negative at the A points and positive at the B points, i.e., $\kappa < 0$.

The Pyragas curve has to lie below the Hopf curve for $\mu < 0$, which means that the slopes at $\mu = 0$ have to obey $\partial_\mu \tau_p > \partial_\mu \tau_H$. This gives

$$\operatorname{tr} K < -\frac{b}{a}\kappa - \frac{\omega}{n\pi}. \tag{5.14}$$

This is the same as condition (5.11), which is no contradiction, since κ has opposite sign. Finally the order of the Hopf points has to be $\tau_n^B \geq \tau_n^A$. A similar discussion as above in the increasing period case shows that there can only be a solution if

$$-\omega \frac{2n+1}{2n^2} \leq \kappa$$

and that the boundary for $\mathrm{tr}\, K$ is the same as above

$$\mathrm{tr}\, K \geq \frac{\kappa \omega}{\sqrt{\omega^2 - 2\kappa\omega}} \cot\left(\pi n \sqrt{1 - \frac{2\kappa}{\omega}} \right).$$

5.4 Summary of the Results

In summary we have the following conditions for stabilization. The domain of κ depends on whether the period increases or decreases with increasing distance from the bifurcation

$$\kappa > 0, \quad \text{if} \quad c - bd/a > 0 \quad \text{(increasing period)},$$
$$\kappa < 0, \quad \text{if} \quad c - bd/a < 0 \quad \text{(decreasing period)}.$$

In any case the domain of control is bounded by

$$-\omega \frac{2n+1}{2n^2} \leq \kappa \leq \omega \frac{2n-1}{2n^2}, \tag{5.15a}$$

$$\mathrm{tr}\, K \geq \frac{\kappa \omega}{\sqrt{\omega^2 - 2\kappa\omega}} \cot\left(\pi n \sqrt{1 - \frac{2\kappa}{\omega}} \right), \tag{5.15b}$$

$$\mathrm{tr}\, K < -\frac{b}{a}\kappa - \frac{\omega}{n\pi}. \tag{5.15c}$$

Figures 5.1 and 5.2 depict the control domain for the two cases. The dotted line in Fig. 5.1 marks the stability domain of the Pyragas orbit for a finite value of $\mu = -0.005$ calculated with the continuation software KNUT (formerly known as PDDE-CONT [4]. For small values of μ the domain is well approximated by the analytic result.As the two boundary curves (5.15b and c)) intersect in the point $(\kappa, \mathrm{tr}\, K) = (0, -\omega/n\pi)$ stabilization is not possible with symmetric feedback matrices, because these have $\kappa = 0$. It is easy to check that even in the case $\det K \neq 0$ control is not possible with $\kappa = 0$, because the Hopf curves are tangent to the τ-axis at the Pyragas points and do not cross the τ-axis at these points. This includes the result of Chap. 3, where feedback with zero rotation angle does not allow control. This imposes a severe limitation for the experimental applicability, because the case $\kappa = 0$ occurs when one can only measure a single variable and

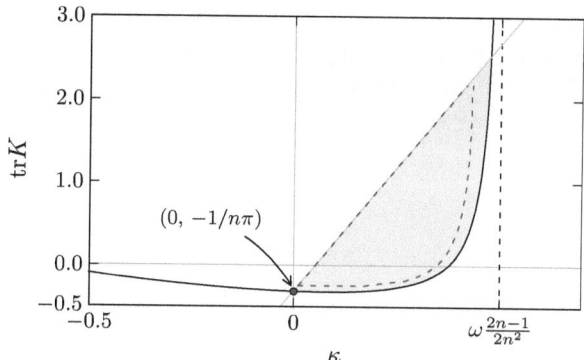

Fig. 5.1 Control domain for the case of increasing period with $\kappa > 0$. The *black* and *gray curve* corresponds to (5.15b and c), respectively. The *dashed vertical line* marks the boundary corresponding to the right boundary in (5.15a). The *gray line* has a slope of $-b/a$. The *dotted line* shows the actual stability domain of the target orbit for $\mu = 0.005$ and is calculated with the continuation software <u>KNUT</u>. Parameters: $\omega = 1$, $n = 1$, $b/a = -6$

Fig. 5.2 Control domain for the case of decreasing period with $\kappa < 0$. The *dashed* and *solid curves* corresponds to (5.15b and c) , respectively. The *dotted line* marks the boundary corresponding to the left boundary in (5.15a). The *black line* has a slope of $-b/a$. Parameters: $\omega = 1$, $n = 1$, $b/a = 6$

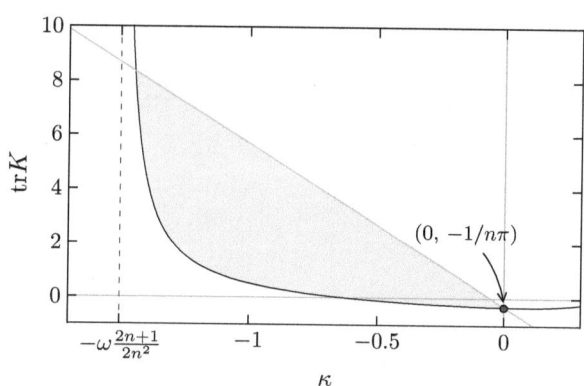

apply the control signal to the dynamic equation of the same variable. Due to the importance of this situation we will in the next section discuss a method to overcome this restriction.

We can find some conditions which ensure a non-empty control domain. The right hand side of (5.15b) is a convex function of κ and has a slope of $-1/2\pi n$ at the intersection point. The right hand side of (5.15c) has a slope of $-b/a$. The following conditions thus lead to a non-empty control domain

$$-\frac{b}{a} > -\frac{1}{2\pi n} \quad \text{for the increasing period case } (\kappa > 0),$$

$$-\frac{b}{a} < -\frac{1}{2\pi n} \quad \text{for the decreasing period case } (\kappa < 0).$$

From these two equations we can see that in the decreasing period case stabilization is only possible for hard springs ($b > 0$). Whereas in the increasing period case we are able to stabilize soft springs ($b < 0$) as well as weakly hard springs ($0 < b < 1/2\pi n$).

References

1. B. Fiedler, S. Yanchuk, V. Flunkert, P. Hövel, H.J Wünsche, E. Schöll, Delay stabilization of rotating waves near fold bifurcation and application to all-optical control of a semiconductor laser, Phys. Rev. E **77**, 066207 (2008)
2. V. Flunkert, E. Schöll, Towards easier realization of time-delayed feedback control of odd-number orbits. Phys. Rev. E (2011, in press)
3. B. Fiedler, V. Flunkert, M. Georgi, P. Hövel, E. Schöll, Refuting the odd number limitation of time-delayed feedback control, Phys. Rev. Lett **98**, 114101 (2007)
4. R. Szalai, Knut, A continuation and bifurcation software for delay-differential equations (2009)

Chapter 6
Stabilization with Symmetric Feedback Matrices

As discussed in the last section, stabilization is not possible with symmetric feedback matrices ($\kappa = 0$). This case, however, is important for experiments as it corresponds to the situation, where one measures a variable and applies the control signal to the dynamical equation of the same variable. We will now discuss a method to overcome this problem, i.e., to stabilize the UPO with symmetric feedback matrices [1].

6.1 Model and Analysis

Consider the normal form model as given by (5.2) with an additional latency time δ in the feedback

$$\frac{d}{dt}\begin{pmatrix} x \\ y \end{pmatrix} = \begin{bmatrix} d\mu + ar^2 & -(\omega + c\mu + br^2) \\ \omega + c\mu + br^2 & d\mu + ar^2 \end{bmatrix}\begin{pmatrix} x \\ y \end{pmatrix}$$
$$+ \begin{bmatrix} K_{11} & K_{12} \\ K_{21} & K_{22} \end{bmatrix}\begin{pmatrix} x_{\tau+\delta} - x_\delta \\ y_{\tau+\delta} - y_\delta \end{pmatrix}. \tag{6.1}$$

Proceeding as above we find the characteristic equation for the eigenvalues of the FP

$$\chi(\eta) = (d\mu - \eta)^2 + \mathrm{tr}K(d\mu - \eta)F(\eta) + \det K\, F(\eta)^2$$
$$+ (\omega + c\mu)^2 + \kappa(\omega + c\mu)F(\eta)$$

with $F(\eta) = e^{-\eta(\tau+\delta)} - e^{-\eta\delta}$ in this case. We consider $\det K = 0$ and $\kappa = 0$, which was not controllable before

$$\chi(\eta) = (d\mu - \eta)^2 + \mathrm{tr}\,K(d\mu - \eta)F(\eta) + (\omega + c\mu)^2. \tag{6.2}$$

V. Flunkert, *Delay-Coupled Complex Systems*, Springer Theses,
DOI: 10.1007/978-3-642-20250-6_6, © Springer-Verlag Berlin Heidelberg 2011

(1) *Location of Hopf points*—To find the Hopf points, we evaluate the real and imaginary part of $0 = \chi(\eta)$ at $\mu = 0$, $\eta = i\Omega$

$$0 = -\Omega^2 + \omega^2 + \Omega \operatorname{tr} K[\sin(\Omega\delta) - \sin(\Omega\delta + \Omega\tau)], \qquad (6.3a)$$

$$0 = \Omega \operatorname{tr} K[\cos(\Omega\delta) - \cos(\Omega\delta + \Omega\tau)]. \qquad (6.3b)$$

The second equation yields

$$\pm\Omega\delta + 2\pi n = \Omega\delta + \Omega\tau.$$

The "+"-sign gives the Hopf A series

$$\tau_n^A = \frac{2\pi n}{\omega}, \qquad \Omega^A = \omega.$$

The "−"-sign gives

$$\tau_n^B = \frac{2\pi n}{\Omega^B} - 2\delta.$$

Inserting this expression into (6.3a) gives a transcendental equation for Ω^B

$$0 = f(\Omega) := \Omega^2 - \omega^2 - 2\operatorname{tr} K\,\Omega\sin(\Omega\delta). \qquad (6.4)$$

Although we cannot explicitly obtain $\Omega^B(\delta)$, we can obtain an explicit parametric representation. To do this we introduce a curve parameter $\psi = \Omega\delta$. Solving $0 = f(\Omega)$ for Ω then gives the parametric solution

$$\Omega^B = \operatorname{tr} K \sin\psi + \sqrt{\omega^2 + (\operatorname{tr} K)^2\sin^2\psi}, \qquad (6.5a)$$

$$\delta = \frac{\psi}{\Omega^B} = \frac{\psi}{\operatorname{tr} K \sin\psi + \sqrt{\omega^2 + (\operatorname{tr} K)^2\sin^2\psi}}. \qquad (6.5b)$$

The Ω^B values lie in in the interval

$$\Omega^B \in [\Omega_{min}, \Omega_{max}], \quad \text{with}$$

$$\Omega_{min} = -\operatorname{tr} K + \sqrt{\omega^2 + (\operatorname{tr} K)^2},$$

$$\Omega_{max} = \operatorname{tr} K + \sqrt{\omega^2 + (\operatorname{tr} K)^2}.$$

Figure 6.1 depicts the solutions Ω^B vs. δ. At the special points

$$\delta_k^* = \frac{\pi}{\omega}k, \qquad (k \in \mathbb{N}_0)$$

Fig. 6.1 Solutions Ω^B of (6.4) vs latency δ. The blue and gray lines correspond to tr$K = 0.1 > 0$ and tr$K = -0.1 < 0$, respectively. Parameters: $\omega = 1$

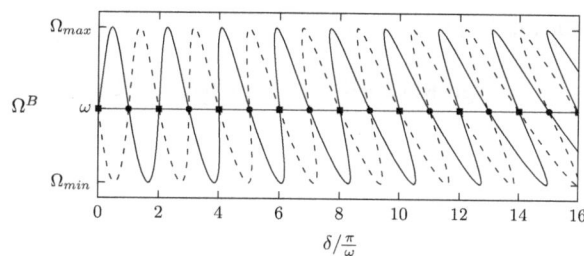

$\Omega = \omega$ is always a solution of (6.4) and some of the Hopf B points lie on the Pyragas points. Note that the labeling of Hopf A and B points is different in this case, i.e.,

$$\tau_n^A = \frac{2\pi n}{\omega} = \frac{2\pi(n+k)}{\omega} - 2\delta_k^* = \tau_{n+k}^B. \tag{6.6}$$

For simplicity we will now consider the case, where (6.4) has a single positive solution, i.e., we consider δ and $|\text{tr}K|$ small enough such that the curve in Fig. 6.1 does not fold back. Implicit differentiation of (6.4) respect to δ gives the slope of the Ω^B curve at $\delta = \delta_k^*$

$$\partial_\delta \Omega^B \Big|_{\delta = \delta_k^*} = \frac{\text{tr}\,K\omega}{(-1)^k - \omega\delta_k^*\text{tr}\,K}.$$

A condition ensuring that a single solution can be found by requiring that the signs of the slopes at the δ_k^* values alternate with k. This is the case if

$$|\text{tr}\,K|\delta < 1/\omega. \tag{6.7}$$

(2) *Crossing direction of Hopf eigenvalue pair*—To calculate the crossing direction of the Hopf eigenvalue pair, we differentiate (6.2) implicitly with respect to τ. This expression evaluated at $\mu = 0$, $\eta = i\Omega$ then gives the crossing direction

$$\text{sgn Re}(\partial_\tau \eta) = \text{sgn}\Big[-(\text{tr}K)^2 + (\text{tr}K)^2\cos(\Omega\tau) + \Omega(\text{tr}K)^2\delta\sin(\Omega\tau)$$
$$- 2\text{tr}K\Omega\sin(\Omega\tau + \Omega\delta)\Big].$$

For the Hopf A series this gives

$$\text{sgn Re}(\partial_\tau \eta)\Big|_A = -\text{sgn}\Big[\text{tr}\,K\sin(\omega\delta)\Big].$$

For the Hopf B series we find

$$
\begin{aligned}
\operatorname{sgn}\,\operatorname{Re}(\partial_\tau\eta)\Big|_B &= -\,\operatorname{sgn}\Big[\operatorname{tr}K\sin(\Omega^B\delta)\Big]\\
&\quad\cdot\operatorname{sgn}\Big[-\Omega+\Omega\,\operatorname{tr}K\delta\cos(\Omega^B\delta)+\operatorname{tr}K\sin(\Omega^B\delta)\Big]\\
&= -\,\operatorname{sgn}\Big[\operatorname{tr}K\sin(\Omega^B\delta)\Big]\cdot\operatorname{sgn}\Big[-f'(\Omega^B)\Big]\\
&= \operatorname{sgn}\Big[(\Omega^B)^2-\omega^2\Big]\cdot\operatorname{sgn}f'(\Omega^B)
\end{aligned}
$$

From (6.4) we find that $f(0)=-\omega^2$ and $\lim_{\Omega\to\infty}f(\Omega)=\infty$, and because we consider the case of a single positive solution $0=f(\Omega^B)$ the slope $f'(\Omega^B)$ is positive. The single solution Ω^B oscillates with increasing δ around ω (see Fig. 6.1) and is larger than ω if $\operatorname{tr}K>0$ and $\delta\in(\delta_k^*,\delta_{k+1}^*)$ with even k or if $\operatorname{tr}K<0$ and $\delta\in(\delta_k^*,\delta_{k+1}^*)$ with odd k. Thus the crossing direction is in fact opposite to that of the A series, i.e.,

$$
\operatorname{sgn}\,\operatorname{Re}(\partial_\tau\eta)\Big|_B=-\operatorname{sgn}\,\operatorname{Re}(\partial_\tau\eta)\Big|_A=\operatorname{sgn}[\operatorname{tr}K\sin(\omega\delta)]\tag{6.8}
$$

(3) *Slope of Hopf and Pyragas curve*—The slope of Hopf curve at the Pyragas points can be calculated by implicit differentiation of the characteristic equation with respect to μ. Evaluated at the Pyragas points we find

$$
\partial_\mu\tau_H=\frac{-2n\pi\,\operatorname{tr}K(c-d\cot(\omega\delta))+2d\omega/\sin(\omega\delta)}{\operatorname{tr}K\omega^2}.\tag{6.9}
$$

6.2 Increasing Period Case

For $-(c-bd/a)<0$ the period of the orbit increases with increasing distance from the bifurcation and the Pyragas curve emanates to the upper left from the Pyragas point. For stabilization we need a (2)-region above and a (0)-region below the nth Pyragas point. This means the eigenvalue crossing direction has to be positive at the A points and negative at the B points, i.e.,

$$
\operatorname{tr}K\sin(\omega\delta)<0.\tag{6.10}
$$

The Pyragas curve has to lie above the Hopf curve for $\mu<0$, i.e., the slopes at $\mu=0$ have to obey $\partial_\mu\tau_p<\partial_\mu\tau_H$, This gives

$$
\frac{b}{a}<\cot(\omega\delta)+\frac{\omega}{\pi n\,\operatorname{tr}K\sin(\omega\delta)},
$$

which can be written as

$$\operatorname{tr} K \left(\frac{b}{a} \sin(\omega\delta) - \cos(\omega\delta) \right) > \frac{\omega}{\pi n}, \tag{6.11}$$

taking into account (6.10).

There is at most one Hopf B point between two successive A points, because

$$\Delta\tau^B = \frac{2\pi}{\Omega^B} > \frac{2\pi}{\omega} = \Delta\tau^A.$$

To have a (0)-region below the Pyragas point, there has to be exactly one such point in between, to compensate for the increase in the number of unstable dimensions at the A points. The Hopf B points start at $\tau_0^B = -2\delta$ and thus the first B point with positive τ^B is given by $\tau_{\tilde{k}}^B$, where $\tilde{k}$ is the smallest integer with

$$\tilde{k} \cdot \Delta\tau^B > 2\delta,$$

i.e.,

$$\tilde{k} = \left\lceil \frac{2\delta}{\Delta\tau^B} \right\rceil = \left\lceil \delta \frac{\Omega^B}{\pi} \right\rceil.$$

Here, $\lceil \cdot \rceil$ denotes the ceiling function. With this index the Hopf point ordering can be written as

$$\tau_{\tilde{k}}^B < \tau_1^A < \tau_{\tilde{k}+1}^B < \tau_2^A < \cdots < \tau_{\tilde{k}+n-1}^B < \tau_n^A.$$

It is sufficient to require $\tau_{\tilde{k}+n-1}^B < \tau_n^A$, because this condition is the strictest. Thus we have

$$\left\lceil \delta \frac{\Omega^B}{\pi} \right\rceil - \delta \frac{\Omega^B}{\pi} - 1 < n \left(\frac{\Omega^B}{\omega} - 1 \right).$$

Using the parametric representation with $\psi = \Omega^B \delta$ (6.5) gives

$$X(\psi) := \frac{\omega}{n} \left(\left\lceil \frac{1}{\pi}\psi \right\rceil - \frac{1}{\pi}\psi - 1 \right) + \omega < \Omega^B.$$

Inserting $\Omega^B(\psi, \operatorname{tr} K)$ from (6.5) and solving for $\operatorname{tr} K$ then gives the boundary curve in parametric form

$$|\operatorname{tr} K(\psi)| < \left| \frac{X(\psi)^2 - \omega^2}{2X(\psi) \sin\psi} \right|, \tag{6.12a}$$

$$\delta(\psi) = \frac{\psi}{X(\psi)}. \tag{6.12b}$$

Figure 6.2 shows the domain of control in the $(\delta, \operatorname{tr} K)$-plane.

6.3 Decreasing Period Case

For $-(c - bd/a) > 0$ the period of the orbit decreases with increasing distance from the bifurcation and the Pyragas curve emanates to the lower left from the Pyragas point. For stabilization we need a (0)-region above and a (2)-region below the emanating Pyragas point. This means the eigenvalue crossing directions has to be negative at the A points and positive at the B points, i.e.,

$$\operatorname{tr} K \sin(\omega\delta) > 0. \tag{6.13}$$

The Pyragas curve has to lie below the Hopf curve for $\mu < 0$, i.e., the slopes at $\mu = 0$ have to obey $\partial_\mu \tau_p > \partial_\mu \tau_H$. This gives

$$\frac{b}{a} > \cot(\omega\delta) + \frac{\omega}{\operatorname{tr} K \pi n \sin(\omega\delta)}.$$

This finally gives the same condition as in the increasing period case (6.11)

$$\operatorname{tr} K \left(\frac{b}{a} \sin(\omega\delta) - \cos(\omega\delta) \right) > \frac{\omega}{\pi n}. \tag{6.14}$$

There is at least one Hopf B point between two successive A points, because

$$\Delta\tau^B = \frac{2\pi}{\Omega^B} < \frac{2\pi}{\omega} = \Delta\tau^A.$$

To have a (2)-region below the Pyragas point, there has to be exactly one such point in between. The Hopf B points start at $\tau_0^B = -2\delta$ and again the first B point with positive τ^B is given by $\tau_{\tilde{k}}^B$, where $\tilde{k}$ is given as above by

$$\tilde{k} = \left\lceil \frac{2\delta}{\Delta\tau^B} \right\rceil = \left\lceil \delta \frac{\Omega^B}{\pi} \right\rceil.$$

The Hopf point ordering is then given by

$$\tau_{\tilde{k}}^B < \tau_1^A < \tau_{\tilde{k}+1}^B < \tau_2^A < \cdots < \tau_n^A < \tau_{\tilde{k}+n}^B.$$

It is sufficient to require $\tau_n^A < \tau_{\tilde{k}+n}^B$, which yields

$$\left\lceil \delta \frac{\Omega^B}{\pi} \right\rceil - \delta \frac{\Omega^B}{\pi} > n \left(\frac{\Omega^B}{\omega} - 1 \right).$$

Using again the parametric representation (6.5) we obtain

$$Y(\psi) := \frac{\omega}{n} \left(\left\lceil \frac{1}{\pi} \psi \right\rceil - \frac{1}{\pi} \psi \right) + \omega > \Omega^B$$

and with $\Omega^B(\psi, \operatorname{tr} K)$ from (6.5) the boundary curve in parametric form

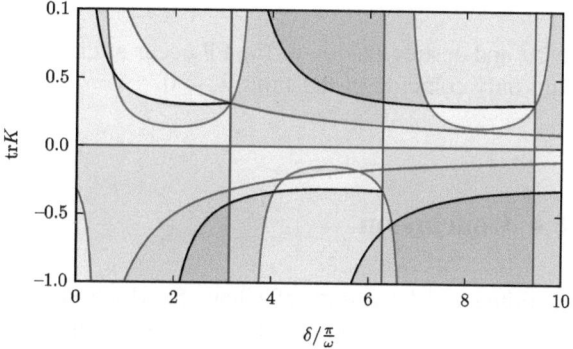

Fig. 6.2 Control domain in the $(\delta, \mathrm{tr}\,K)$-plane. The yellow regions are the regions of control. The other colored regions depict domains, where one or more of the control conditions is violated: *blue*—(6.10), *red*—(6.11), *green*—(6.7), *gray*—(6.12a). Parameters: $\omega = 1$, $b/a = -2$, $c = 4$, $d = 1$, $n = 1$

Fig. 6.3 Control domain for the decreasing period case in the $(\delta, \mathrm{tr}\,K)$-plane. The yellow regions are the regions of control. The other colored regions depict domains, where one or more of the control conditions is violated: *blue*—(6.13), *red*—(6.14), *green*—(6.7), *gray*—(6.15a). Parameters: $\omega = 1$, $b/a = 2$, $c = -4$, $d = 1$, $n = 1$

$$|\mathrm{tr}\,K(\psi)| < \left| \frac{Y(\psi)^2 - \omega^2}{2Y(\psi)\sin\psi} \right|, \tag{6.15a}$$

$$\delta(\psi) = \frac{\psi}{Y(\psi)}. \tag{6.15b}$$

Note that since we only consider the case where (6.4) has a single solution (see (6.7)) all the conditions we constructed are sufficient but not necessary. The actual domains of control in Figs. 6.2 and 6.3 may in fact be larger. Figure 6.4 depicts for the increasing period case the analytically found control domain and the actual stability boundary (dotted blue line) of the target orbit in the increasing period case for $\mu = 0.005$ calculated with KNUT. Our simplifying restriction (6.11) (green line) to a single solution of (6.4) is not necessary for this set of parameters, i.e., the additional solutions of (6.4) do not prevent successful control. For other parameters this may not be the case and then (6.11) ensures successful control. Note that in Fig. 6.4 the actual stability domains of the orbit are shifted with respect to the analytically found domains, due to the finite value of $\mu = 0.005$. This is the same effect as occurs in Fig. 3.10, where for finite values of λ the stabilization of the

Fig. 6.4 Control domain
(*dotted blue line*) of the target
orbit calculated with KNUT.
The other lines correspond to
the analytically found
boundaries of control as in
Fig. 6.2. Parameters: $\omega = 1$,
$b/a = -2$, $c = 4$, $d = 1$,
$n = 1$, $\mu = 0.005$

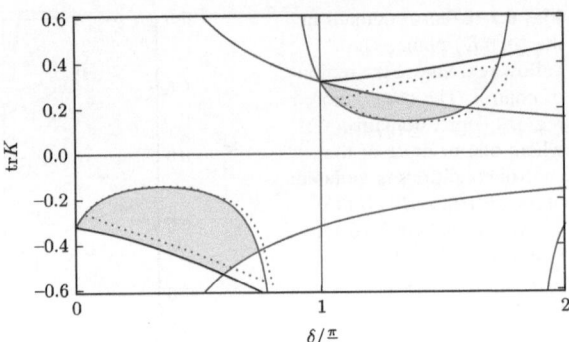

orbit and destabilization of the FP occur at different values of the control strengths
and only coincide in the limit $\lambda \to 0$.

6.4 Conclusion

Building on the results of Chap. 5, which showed that feedback matrices with
$\det K = 0$ are experimentally relevant, we have in this section considered symmetric feedback matrices with $\det K = 0$. These feedback matrices correspond to
the case, when the input and output variable for the control are the same. In the
previous Chaps. 3,4, and 5 we saw that stabilization was not possible with symmetric feedback matrices. In this section we showed that stabilization is possible,
if we introduce an additional latency time to the Pyragas control. As in Chap. 5, we
needed to distinguish the increasing period case and the decreasing period case,
which are both controllable.

Latencies arise naturally in control loops due to the finite signal processing and
propagation speed in the loop. For instance, in a laser, where the Pyragas control
can be realized with a Fabry-Perot resonator, the distance between the resonator
and the laser introduces a latency in the control signal [2]. In many cases it is easy
to tune the overall latency of a control loop or to deliberately introduce an additional latency in the loop, which can be varied. This makes the overall control loop
latency an accessible control parameter.

The effects of latencies in Pyragas control loops have previously been studied
[3–6]. It has been observed that in general such latencies result in shifted and
slightly deformed control regions. Similar effects arise from phase-dependent
control amplitudes and in the above references it has been shown that control
phase and latency can have a qualitatively similar effect or for proper tuning
compensate each other. This intuitively explains why a latency in the feedback can
stabilize odd-number orbits with symmetric feedback matrices.

References

1. V. Flunkert, E. Schöll, Towards easier realization of time-delayed feedback control of odd-number orbits. Phys. Rev. E (2011, in press)
2. T. Dahms, P. Hövel, E. Schöll, Stabilizing continuous-wave output in semiconductor lasers by time-delayed feedback. Phys. Rev. E **78**, 056213 (2008)
3. P. Hövel, J.E.S Socolar, Stability domains for time-delay feedback control with latency. Phys. Rev. E **68**, 036206 (2003)
4. P. Hövel, E. Schöll, Control of unstable steady states by time-delayed feedback methods. Phys. Rev. E **72**, 046203 (2005)
5. S. Schikora, P. Hövel, H.J Wünsche, E. Schöll, F. Henneberger, All-optical noninvasive control of unstable steady states in a semiconductor laser. Phys. Rev. Lett. **97**, 213902 (2006)
6. T. Dahms, P. Hövel, E. Schöll, Control of unstable steady states by extended time-delayed feedback. Phys. Rev. E **76**, 056201 (2007)

Chapter 7
Application to Laser Systems

In this section we will look at the stabilization of odd-number orbits in laser systems. Lasers subject to delayed feedback have been studied since the seminal paper of Lang and Kobayashi [1]. Various delayed feedback methods have since then been used,[1] such as all-optical feedback [1], phase-conjugate feedback [2], optoelectronic feedback [3–5], polarization rotated feedback [6–9] and filtered feedback [10].

These lasers show very interesting dynamics [11, 12] including complicated bifurcation scenarios [13, 14] and chaos [5, 15, 16] and are of immense practical importance due to their applications in telecommunication. Delayed feedback schemes can for example be used to suppress noise [17–19] and to control chaos [20–23].

Pyragas control has also been successfully implemented experimentally in lasers and has been used for the noninvasive stabilization of FPs and POs [23–30]. Experimentally, the Pyragas feedback can be realized all optically in a natural way by coupling the laser to a Michelson interferometer [31] or an external Fabry-Perot [29] resonator. When the phase conditions in the resonator are chosen appropriately the laser receives optical feedback of the form $Ke^{i\phi}[E(t) - E(t - \tau)]$, where E is the complex electric field of the laser light and $Ke^{i\phi}$ is the coupling strengths including a possible phase shift of the overall signal. Note that in this form we neglect multiple reflections in the resonator. This assumption is valid for small values of K, since in this case the terms corresponding to multiple reflections $K^2 \ll K$ are very small. For larger values of K these multiple reflections cannot be neglected and one obtains the form of extended time-delayed feedback [32, 33]. Another possibility to apply the feedback is to use optoelectronic coupling [34, 35], which we will discuss in more detail in Sect. 7.2.

---

[1] Due to the vast amount of literature in this field it is impossible to give a comprehensive list of references.

V. Flunkert, *Delay-Coupled Complex Systems*, Springer Theses,
DOI: 10.1007/978-3-642-20250-6_7, © Springer-Verlag Berlin Heidelberg 2011

7.1 Stabilization of an Anti-Mode

As a first example of time-delayed feedback control in lasers, we will stabilize an anti-mode of the Lang-Kobayashi laser system. Consider the dimensionless Lang-Kobayashi equations

$$\frac{d}{dt}E = \frac{1}{2}(1 + i\alpha)nE + KE(t - \sigma) + b[E(t - \tau) - E(t)],$$

$$T\frac{d}{dt}n = p - n - (1 + n)|E|^2.$$

Here, E is the complex electric field amplitude, n is the carrier density (in excess of the threshold carrier density), α is the alpha or linewidth enhancement factor, p is the dimensionless pump current, which is zero at threshold, and $T = T_e/T_p$ is the time scale parameter, where T_e and T_p are the carrier and photon lifetimes, respectively. The term $KE(t - \sigma)$ corresponds to the feedback term of the uncontrolled Lang-Kobayashi system, which induces the modes and antimodes, and $b[E(t - \tau) - E(t)]$ is the Pyragas control term. See Chap. 11 for a more detailed introduction and discussion of the model.

The RW solutions $E = A\,e^{i\omega t}$ obey

$$0 = \frac{1}{2}n + K\cos(\omega\sigma) + b_0[\cos(\beta - \omega\tau) - \cos\beta],$$

$$\omega = \frac{1}{2}\alpha n - K\sin(\omega\sigma) + b_0[\sin(\beta - \omega\tau) - \sin\beta],$$

$$0 = p - n - (1 + n)A^2.$$

Eliminating n in the first two equations we find, similar to Sect. 3.4, a transcendental frequency equation

$$\omega = -K\Big[\sin(\omega\sigma) + \alpha\cos(\omega\sigma)\Big]$$
$$+ b_0\Big[\sin(\beta - \omega\tau) - \alpha\cos(\beta - \omega\tau) - \sin\beta + \alpha\cos\beta\Big]. \qquad (7.1)$$

Using the ansatz

$$E(t) = A(t)e^{i\phi(t)} \quad \text{with} \quad \dot{E} = (\dot{A} + i\dot{\phi}A)e^{i\phi}$$

we can rewrite the Lang-Kobayashi equations in amplitude A and phase ϕ

$$\frac{d}{dt}A = \frac{1}{2}nA + KA_\sigma\cos(\phi_\sigma - \phi) + b_0[A_\tau\cos(\beta + \phi_\tau - \phi) - A\cos\beta],$$

$$\frac{d}{dt}\phi = \frac{\alpha}{2}n + K\frac{A_\sigma}{A}\sin(\phi_\sigma - \phi) + b_0\left[\frac{A_\tau}{A}\sin(\beta + \phi_\tau - \phi) - \sin\beta\right],$$

$$T\frac{d}{dt}n = p - n - (1 + n)A^2.$$

We now linearize the equations around the external cavity mode (see Sect. 11.3)

$$\frac{d}{dt}X(t) = M_1 X(t) + M_2 X(t - \sigma) + M_3 X(t - \tau)$$

with $X := (A, \phi, n)$ and

$$M_1 = \begin{bmatrix} \frac{1}{2}n - b_0 \cos\beta & A[b_0 \sin\beta - K \sin(\omega\sigma)] & \frac{1}{2}A \\ \frac{1}{A}[-b_0 \sin\beta + K \sin(\omega\sigma)] & -b_0 \cos\beta - K \cos(\omega\sigma) & \frac{1}{2}\alpha \\ -\frac{1}{T}2A(1+n) & 0 & -\frac{1}{T}(1+A^2) \end{bmatrix},$$

$$M_2 = \begin{bmatrix} K \cos(\omega\sigma) & AK \sin(\omega\sigma) & 0 \\ -\frac{K}{A} \sin(\omega\sigma) & K \cos(\omega\sigma) & 0 \\ 0 & 0 & 0 \end{bmatrix},$$

$$M_3 = \begin{bmatrix} b_0 \cos\beta & -Ab_0 \sin\beta & 0 \\ \frac{b_0}{A} \sin\beta & b_0 \cos\beta & 0 \\ 0 & 0 & 0 \end{bmatrix}.$$

With the same procedure as in Chap. 4 we can find the transcritical bifurcation. We expand the characteristic equation

$$\chi(z) := \det\left[-z + M_1 + M_2 e^{-z\sigma} + M_3 e^{-z\tau}\right]$$

up to second order in z eliminate the Goldstone mode by factorizing out the common factor z and set $z \stackrel{!}{=} 0$. This gives the control gain at the bifurcation as a function of β

$$b_0(\beta) = \frac{a}{b \cos\beta + c \sin\beta},$$

where the coefficients are given by

$$a = -\left(-n + A^2(2+n) - 2(1+A^2)K \cos(\omega\sigma)\right)\left(1 + K\sigma \cos(\omega\sigma)\right)$$
$$+ 2A^2 K(1+n)\alpha\sigma \sin(\omega\sigma),$$
$$b = \tau\left(-n + A^2(2+n) - 2(1+A^2)K \cos(\omega\sigma)\right),$$
$$c = 2A^2(1+n)\alpha\tau.$$

Figure 7.1(b) depicts the solutions of the transcendental frequency (7.1) as a function of the feedback strengths K for the uncontrolled laser ($\sigma = 0$). As the feedback strength is increased, stable modes (solid lines) and unstable antimodes (dashed lines) are created in saddle-node bifurcations (see Sect. 11.3). Panel (a) shows a zoom into the saddle-node bifurcation marked with SN in panel (b). The blue cross marks the target orbit which we aim to stabilize. Choosing the control

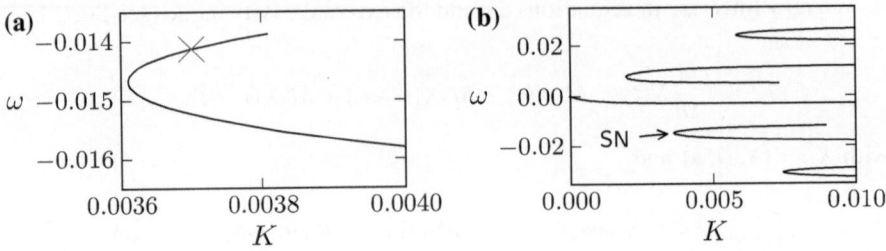

Fig. 7.1 Fold bifurcations of modes (*solid*) and anti-modes (*dashed*) in the (K, ω)-plane for the uncontrolled ($b_0 = 0$) LK system. Panel (**a**) shows a zoom into the saddle-node bifurcation marked with *SN* in panel (**b**). The blue cross marks the target state. Parameters: $\alpha = 4$, $T = 200$, $p = 1$, $\sigma = 400$

parameters as $\sigma = 400$, $b_0 = 0.002$, and $\beta = \pi/2$ leads to stabilization of this mode.

In Fig. 7.2 the solutions of the transcendental frequency equations are given by the intersection of the straight line with the black curve for the uncontrolled and the with the blue curve for the controlled system. For this value of the control parameters the anti-mode of the uncontrolled system becomes a mode of the controlled system, i.e., is stabilized noninvasively.

It might seem strange to stabilize an anti-mode by time-delayed feedback control, which is generated by another delay term in the first place. However, this is only a simple example, where Pyragas control of a laser can stabilize an unstable orbit born in a saddle-node bifurcation. A different system of two coupled lasers, which also exhibits a saddle-node bifurcation, has, for instance, been studied in [28]. In this example it is also possible to stabilize the unstable branch.

7.2 Stabilization of Intensity Pulsations with Optoelectronic Feedback

In the following we will discuss different feedback schemes with the aim of stabilizing a subcritical Hopf bifurcation in a laser. Consider the following laser model [36] describing a laser with a passive dispersive reflector

$$\frac{d}{dt}\rho = n\rho,$$

$$T\frac{d}{dt}n = p - n - (1 + n)k_\mu(n)\rho,$$

where ρ is the intensity of the laser and

$$k_\mu(n) = K_\mu + \frac{AW^2}{4(n - \mu)^2 + W^2}$$

Fig. 7.2 Transcendental
frequency equation for
controlled (*blue curve*) an
uncontrolled (*black curve*)
system. Panel (**b**) shows a
zoom into the box indicated
in panel (**a**). Parameters:
$\alpha = 4$, $\sigma = 400$, $K = 0.0037$,
$b_0 = 0.002$, $\beta = \pi/2$,
$\tau = 2\pi/\omega$

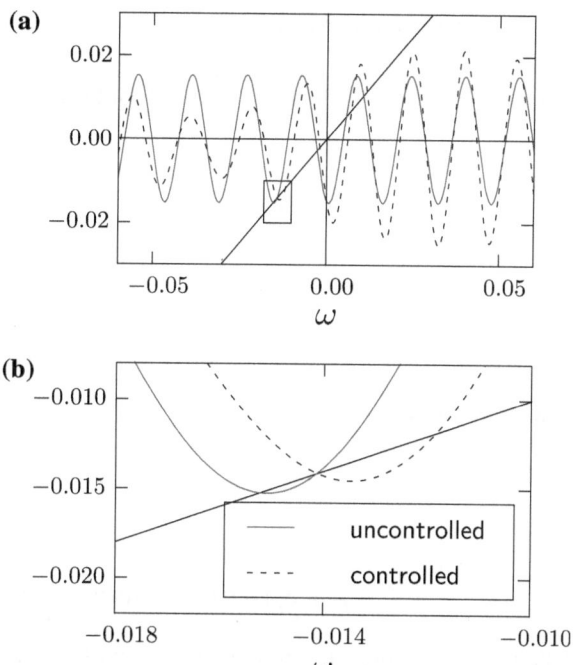

is a Lorentzian function with height $A = 1$ and width $W = 0.02$ describing the
behavior of the reflector. We treat the position μ of the Lorentzian as a
bifurcation parameter and the parameter K_μ is determined by the condition
$k_\mu(0) = 1$ [36]

$$K_\mu = 1 - \frac{AW^2}{4\mu^2 + W^2}.$$

We use this more complicated model, since it exhibits undamped relaxation
oscillations, and in particular a subcritical Hopf bifurcation.

Suppose there is a Hopf bifurcation at $\mu = \mu_*$. To apply the discussion from the
last section, we need to bring the laser equations close to this Hopf bifurcation
(small $\Delta\mu := \mu - \mu_*$) into the normal form

$$\frac{d}{dt}x = \left[d\Delta\mu + a(x^2 + y^2)\right]x - \left[\omega + c\Delta\mu + b(x^2 + y^2)\right]y,$$

$$\frac{d}{dt}y = \left[\omega + c\Delta\mu + b(x^2 + y^2)\right]x + \left[d\Delta\mu + a(x^2 + y^2)\right]y,$$

i.e., we need to find the coefficients a, b, c, and d.

7.2.1 Normal Form Analysis

Let us first discuss the location of the Hopf bifurcation. The Jacobian of the system is given by

$$J = \begin{bmatrix} n & \rho \\ -\frac{1}{T}(1+n)k_\mu(n) & -\frac{1}{T}[1+\rho k_\mu(n) + \rho(1+n)k_\mu'(n)] \end{bmatrix}.$$

At the lasing FP $(\rho, n) = (p, 0)$ the Jacobian is then

$$J = \begin{bmatrix} 0 & p \\ -\frac{1}{T} & -\frac{1}{T}[1 + p + p k_\mu'(0)] \end{bmatrix}.$$

The eigenvalues are

$$\lambda_\pm = -\gamma \pm i\sqrt{\omega^2 - \gamma^2},$$

with

$$\gamma = \frac{1}{2T}[1 + p + p k_\mu'(0)], \qquad \omega = \sqrt{p/T}.$$

The function k_μ depends on μ, which we will treat as a bifurcation parameter. If for some value μ_*

$$k_{\mu_*}'(0) = -\frac{1+p}{p}$$

then the real part of the eigenvalues vanish $\gamma = 0$ and there is a Hopf bifurcation. The values μ_*, where this happens solve the implicit equation

$$\frac{8AW^2\mu_*}{\left(W^2 + 4\mu_*^2\right)^2} = -\frac{1+p}{p}. \tag{7.2}$$

The Jacobian then has eigenvalues

$$\lambda_\pm = \pm i\omega.$$

From the eigenvalues we can already determine two of the coefficients

$$d = \partial_\mu \text{Re}(\lambda_\pm(\mu))\Big|_{\mu=\mu_*} = -\frac{p}{2T}\partial_\mu k_\mu'(0)\Big|_{\mu=\mu_*}$$

$$= \frac{4pAW^2(12\mu^2 - W^2)}{T(4\mu^2 + W^2)^2}$$

and

$$c = \partial_\mu \text{Im}(\lambda_+(\mu))\Big|_{\mu=\mu_*} = \partial_\mu \sqrt{\omega^2 - \gamma^2}\Big|_{\mu=\mu_*} = \frac{-\gamma \partial_\mu \gamma}{\sqrt{\omega^2 - \gamma^2}}\Big|_{\mu=\mu_*} = 0,$$

where we have used $\gamma = 0$ at $\mu = \mu_*$ in the last equation.
Using the transformation

$$U = \begin{bmatrix} p & p \\ \omega & -\omega \end{bmatrix}, \quad U^{-1} = \begin{bmatrix} \frac{1}{2p} & \frac{1}{2\omega} \\ \frac{1}{2p} & \frac{-1}{2\omega} \end{bmatrix} \tag{7.3}$$

we define the new coordinates x and y according to

$$\begin{pmatrix} x \\ y \end{pmatrix} = U^{-1} \begin{pmatrix} \rho - p \\ n \end{pmatrix}, \quad \begin{pmatrix} \rho \\ n \end{pmatrix} = U \begin{pmatrix} x \\ y \end{pmatrix} + \begin{pmatrix} p \\ 0 \end{pmatrix}.$$

In these new coordinates the dynamical equations are given by

$$\frac{d}{dt} \begin{pmatrix} x \\ y \end{pmatrix} = \begin{bmatrix} d\Delta\mu & -\omega \\ \omega & d\Delta\mu \end{bmatrix} \begin{pmatrix} x \\ y \end{pmatrix} + \begin{pmatrix} f(x,y) \\ g(x,y) \end{pmatrix},$$

where f and g carry the nonlinear terms. With the notation

$$f_{xy} = \frac{\partial^2 f(0,0)}{\partial x \partial y}, \text{ etc.}$$

we can then calculate the other two coefficients

$$a = \frac{1}{16}\left[f_{xxx} + f_{xyy} + g_{xxy} + g_{yyy}\right]$$
$$+ \frac{1}{16\omega}\left[f_{xy}(f_{xx} + f_{yy}) - g_{xy}(g_{xx} + g_{yy}) - f_{xx}g_{xx} + f_{yy}g_{yy}\right]$$
$$= \frac{2 + (2 - k_{\mu_*}^{(2)})p - (3k_{\mu_*}^{(2)} + k_{\mu_*}^{(3)})p^2}{8T^2},$$

and

$$b = \frac{1}{16}\left[g_{xxx} + g_{xyy} - f_{yyy} - f_{xyy}\right]$$
$$+ \frac{1}{48\omega}\left[f_{xx}g_{xy} + f_{xy}g_{yy} - 2(f_{xx}^2 + f_{xy}^2 + g_{xy}^2 + 2g_{yy}^2)\right.$$
$$\left. - 5(f_{yy}^2 + g_{xx}^2 + f_{xx}f_{yy} + g_{xx}g_{yy} + f_{xy}g_{xx} - f_{yy}g_{xy})\right]$$
$$= -\frac{1}{12\omega T^3}\left[(2 - k_{\mu_*}^{(2)})^2 p^3 + T + (2 - k_{\mu_*}^{(2)})(T + 4)p^2 + (T^2 + 2T + 4)p\right].$$

Here, we have used the expansion

$$k_{\mu_*}(n) = 1 - \frac{1+p}{p}n + \frac{k_{\mu_*}^{(2)}}{2!}n^2 + \frac{k_{\mu_*}^{(3)}}{3!}n^3.$$

We now consider a concrete example with the laser parameters shown in
Table 7.1. With these typical values of the parameters we can start to calculate the

Table 7.1 Parameters of the Hopf laser model

Parameter	p	A	W	T
Value	2.0	1.0	0.02	1000

Fig. 7.3 Bifurcation diagram of the Hopf laser. Modified figure from [36] (courtesy V. Tronciu; Reprinted from Opt. Commun. **182**; V.Z. Tronciu, H.J. Wünsche, J. Sieber, K. Schneider and F. Henneberger, *Dynamics of single mode semiconductor lasers with passive dispersive reflectors*, pp. 221, Copyright (2000), with permission from Elsevier)

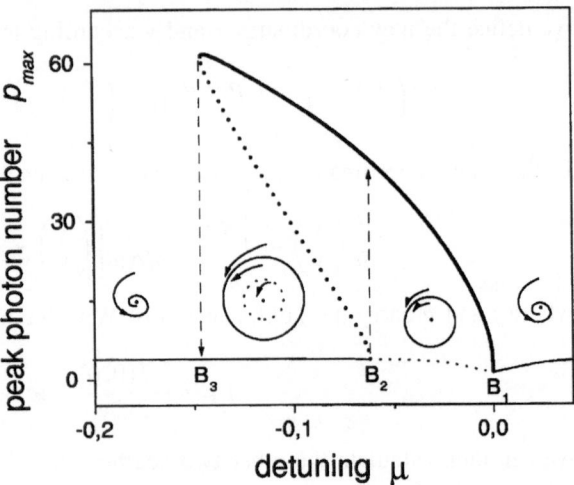

values of μ, where Hopf bifurcations take place, according to (7.2). Solving this equation numerically we find two Hopf bifurcations

$$\mu_{\text{sub}} \approx -4.976 \times 10^{-2}, \qquad \mu_{\text{super}} \approx -7.5 \times 10^{-4},$$

where μ_{sub} is a subcritical bifurcation and μ_{super} is a supercritical bifurcation. The corresponding bifurcation diagram is depicted in Fig. 7.3. Using the laser parameters we can now calculate the Hopf normal form parameters at the sub-critical bifurcation $\mu = \mu_{\text{sub}}$. The approximate values are given in Table 7.2. From the signs of a and d we see that it is indeed a subcritical Hopf bifurcation with the stable FP lying to the left of the bifurcation point ($\mu < \mu_{\text{sub}}$). Furthermore,

$$-(c - bd/a) \approx -8.967 \times 10^{-2} < 0$$

implies that we have the increasing period case.

7.2.2 Optoelectronic Feedback

To stabilize the subcritical Hopf orbit we now consider delayed optoelectronic feedback of the form

Table 7.2 Parameters of the normal form model describing the laser close to the subcritical Hopf bifurcation

Parameter	a	b	c	d	ω
Value	3.074×10^{-3}	-3.214×10^{-3}	0.0	8.576×10^{-2}	4.472×10^{-2}

$$\frac{d}{dt}\rho = n\rho,$$

$$\frac{d}{dt}n = \left[p + \beta(\rho_\tau - \rho) - n - (1+n)k_\mu(n)\rho\right]/T,$$

where the control signal is given by

$$\beta(\rho_\tau - \rho).$$

Such feedback can be realized by measuring the intensity of the laser with a photodiode and modulating the pump current according to the delayed difference signal [4].

To find successful control parameters β we first need to understand what the control term will become after the normal form transformation. Since normal form transformations leave linear terms invariant [37], we only need to take the linear transformation (7.3) into account. This give the control matrix in the normal form coordinates

$$K = U^{-1}\begin{bmatrix} 0 & 0 \\ \frac{\beta}{T} & 0 \end{bmatrix} U = \begin{bmatrix} \frac{1}{2}\omega\beta & \frac{1}{2}\omega\beta \\ -\frac{1}{2}\omega\beta & -\frac{1}{2}\omega\beta \end{bmatrix}. \tag{7.4}$$

The relevant control parameters are then given by

$$\kappa = -\omega\beta, \quad \operatorname{tr} K = 0.$$

In addition, we can find the domain of control in the $(\kappa, \operatorname{tr} K)$-plane from the calculated Hopf normal form coefficients as shown in Fig. 7.4. Using $\operatorname{tr} K = 0$ we can calculate the control interval for κ explicitly

$$\kappa \in \left[-\omega\frac{a}{\pi b}, \ \omega\frac{4n^2 - 1}{8n^2}\right] \approx [0.0136, 0.0168] \tag{7.5}$$

and the corresponding β-interval

$$\beta \in \left[-\frac{4n^2 - 1}{8n^2}, \ \frac{a}{\pi b}\right] \approx [-0.375, -0.304].$$

Figure 7.5 shows the time series of the laser intensity (panel (a)) and the time series of the control signal's amplitude (panel (b)) in the case of successful stabilization.

A very similar feedback can also be realized all-optically by using polarization rotated feedback [6–9]. This is achieved by rotating the polarization axis of the emitted light by $\pi/2$ into the perpendicular orientation and reinjecting this light

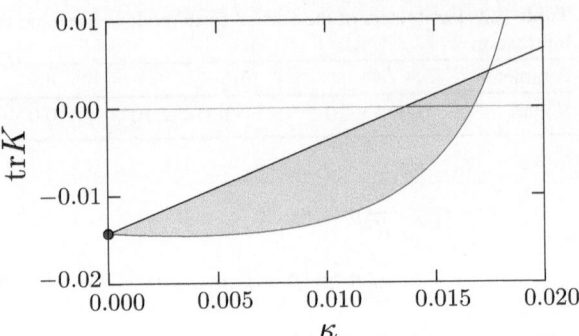

Fig. 7.4 Domain of control for the Hopf normal form with coefficients as in Table 7.2. The intersection of the shaded area with the $\mathrm{tr}K = 0$ line is control interval (7.5)

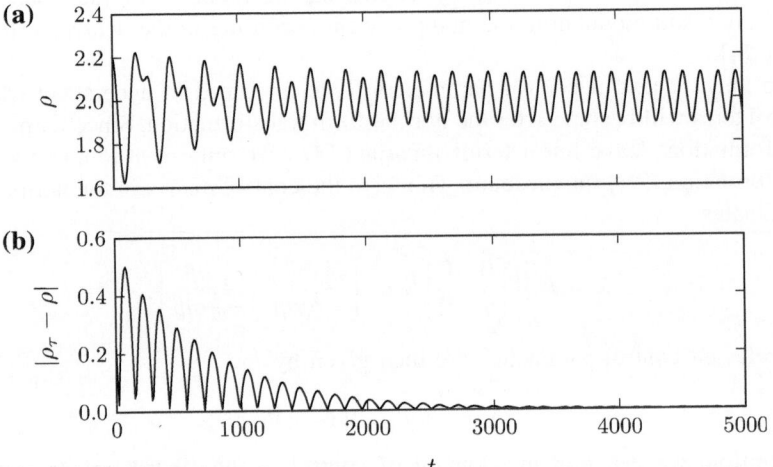

Fig. 7.5 Stabilization of the subcritical Hopf orbit in the laser system. Panel (**a**): Time series of the intensity ρ; Panel (**b**): Time series of the control signal; The laser parameters are as in Table 7.1. The distance to the bifurcation is $\Delta\mu = \mu - \mu_{\mathrm{sub}} = -0.001$ and the control amplitude is chosen as $\beta = -0.3354$ in the control interval

into the laser. Due to the orthogonal polarization this injected signal does not contribute to the lasing mode but is still amplified and thus reduces the inversion. As a result one obtains a feedback of the following form

$$\frac{d}{dt}\rho = n\rho,$$

$$T\frac{d}{dt}n = p - n - (1+n)k_\mu(n)(\rho + \beta\rho_\tau).$$

Through interference it might then be possible to realize the Pyragas control scheme with this method

$$T\frac{d}{dt}n = p - n - (1+n)k_\mu(n)[\rho + \beta(\rho_\tau - \rho)].$$

The advantage would be that the control works even for very high oscillation frequencies beyond the bandwidth of electronic circuits.

7.3 Conclusion

In this section we have shown that it is in principle possible to stabilize odd-number orbits in lasers by means of noninvasive time-delayed feedback control. In particular, we showed that all-optical feedback from a Fabry-Perot resonator can, for precise tuning of the feedback parameters, stabilize an antimode close to the fold bifurcation, where it is generated. Furthermore, optoelectronic feedback can stabilize unstable intensity pulsations close to a subcritical Hopf bifurcation. This is a direct application of the results obtained in Chap. 5.

References

1. R. Lang, K. Kobayashi, External optical feedback effects on semiconductor injection laser properties. IEEE J. Quantum Electron. **16**, 347 (1980)
2. G.P. Agrawal, G.R. Gray, Effect of phase-conjugate feedback on the noise characteristics of semiconductor-lasers. Phys. Rev. A **46**, 5890 (1992)
3. G. Giacomelli, M. Calzavara, F.T. Arecchi, Instabilities in a semiconductor laser with delayed optoelectronic feedback. Opt. Commun. **74**, 97 (1989)
4. H.D.I. Abarbanel, M.B. Kennel, L. Illing, S. Tang, H.F. Chen, J.M. Liu, Synchronization and communication using semiconductor lasers with optoelectronic feedback. IEEE J. Quantum Electron. **37**, 1301 (2001)
5. J. Ohtsubo, Semiconductor Lasers: Stability, Instability and Chaos. (Springer, Berlin, 2005)
6. K. Otsuka, J.L. Chern, High-speed picosecond pulse generation in semiconductor lasers with incoherent optical feedback, Opt. Lett. **16**, 1759 (1991)
7. J. Houlihan, G. Huyet, J. McInerney, Dynamics of a semiconductor laser with incoherent optical feedback. Opt. Commun. **199**, 175 (2001)
8. F. Rogister, A. Locquet, D. Pieroux, M. Sciamanna, O. Deparis, P. Megret, M. Blondel, Secure communication scheme using chaotic laser diodes subject to incoherent optical feedback and incoherent optical injection. Opt. Lett. **26**, 1486 (2001)
9. J.M.S. Solorio, D.W. Sukow, D.R. Hicks, A. Gavrielides, Bifurcations in a semiconductor laser subject to delayed incoherent feedback. Opt. Commun. **214**, 327 (2002)
10. K. Green, B. Krauskopf, Mode structure of a semiconductor laser subject to filtered optical feedback. Opt. Commun. **258**, 243 (2006)
11. T. Heil, I. Fischer, W. Elsäßer, A. Gavrielides, Dynamics of semiconductor lasers subject to delayed optical feedback: The short cavity regime. Phys. Rev. Lett. **87**, 243901 (2001)
12. D.M. Kane, K.A. Shore (eds.), Unlocking Dynamical Diversity: Optical Feedback Effects on Semiconductor Lasers (Wiley VCH, Weinheim, 2005)
13. S. Wieczorek, B. Krauskopf, D. Lenstra, Unifying view of bifurcations in a semiconductor laser subject to optical injection. Opt. Commun. **172**, 279 (1999)
14. V. Rottschäfer, B. Krauskopf, The ECM-backbone of the Lang-Kobayashi equations: A geometric picture. Int. J. Bif. Chaos **17**, 1575 (2007)

15. J. Mørk, B. Tromborg, J. Mark, Chaos in semiconductor lasers with optical feedback-theory and experiment. IEEE J. Quantum Electron. **28**, 93 (1992)
16. I. Fischer, O. Hess, W. Elsäßer, E.O. GÄbel, High-dimensional chaotic dynamics of an external-cavity semiconductor-laser. Phys. Rev. Lett. **73**, 2188 (1994)
17. K. Petermann, Laser Diode Modulation and Noise. (KTK Scientific Publishers, Tokyo, 1988)
18. G.P. Agrawal, N.K. Dutta, Semiconductor Lasers (Van Nostrand Reinhold, New York, 1993)
19. V. Flunkert, E. Schöll, Suppressing noise-induced intensity pulsations in semiconductor lasers by means of time-delayed feedback. Phys. Rev. E **76**, 066202 (2007)
20. C. Simmendinger, O. Hess, Controlling delay-induced chaotic behavior of a semiconductor laser with optical feedback. Phys. Lett. A **216**, 97 (1996)
21. M. Münkel, F. Kaiser, O. Hess, Stabilization of spatiotemporally chaotic semiconductor laser arrays by means of delayed optical feedback. Phys. Rev. E **56**, 3868 (1997)
22. A. Ahlborn, U. Parlitz, Chaos control using notch feedback. Phys. Rev. Lett. **96**, 034102 (2006)
23. S. Schikora, H.J. Wünsche, F. Henneberger, All-optical noninvasive chaos control of a semiconductor laser. Phys. Rev. E **78**, 025202 (2008)
24. D.J. Gauthier, D.W. Sukow, H.M. Concannon, J.E.S. Socolar, Stabilizing unstable periodic orbits in a fast diode resonator using continuous time-delay autosynchronization. Phys. Rev. E **50**, 2343 (1994)
25. V.Z. Tronciu, H.J. Wünsche, M. Wolfrum, M. Radziunas, Semiconductor laser under resonant feedback from a Fabry-Perot: Stability of continuous wave operation, submitted to PRE (2005)
26. V.Z. Tronciu, H.J. Wünsche, M. Wolfrum, M. Radziunas, Semiconductor laser under resonant feedback from a Fabry-Perot: Stability of continuous-wave operation. Phys. Rev. E **73**, 046205 (2006)
27. S. Schikora, P. Hövel, H.J. Wünsche, E. Schöll, F. Henneberger, All-optical noninvasive control of unstable steady states in a semiconductor laser. Phys. Rev. Lett. **97**, 213902 (2006)
28. B. Fiedler, S. Yanchuk, V. Flunkert, P. Hövel, H.J. Wünsche, E. Schöll, Delay stabilization of rotating waves near fold bifurcation and application to all-optical control of a semiconductor laser. Phys. Rev. E **77**, 066207 (2008)
29. T. Dahms, P. Hövel, E. Schöll, Stabilizing continuous-wave output in semiconductor lasers by time-delayed feedback. Phys. Rev. E **78**, 056213 (2008)
30. S. Schikora, H.J. Wünsche, and F. Henneberger, (2010) in preparation
31. C. Simmendinger, M. Münkel, O. Hess, Controlling complex temporal and spatio-temporal dynamics in semiconductor lasers. Chaos, Solitons Fractals **10**, 851 (1999)
32. J.E.S. Socolar, D.W. Sukow, D.J. Gauthier, Stabilizing unstable periodic orbits in fast dynamical systems. Phys. Rev. E **50**, 3245 (1994)
33. A. Chang, J.C. Bienfang, G.M. Hall, J.R. Gardner, D.J. Gauthier, Stabilizing unstable steady states using extended time-delay autosynchronisation. Chaos **8**, 782 (1998)
34. K.E. Callan, L. Illing, Z. Gao, D.J. Gauthier, E. Schöll, Broadband chaos generated by an opto-electronic oscillator. Phys. Rev. Lett. **104**, 113901 (2010)
35. L. Larger, J.M. Dudley, Nonlinear dynamics: Optoelectronic. chaos. Nature **465**, 41 (2010)
36. V.Z. Tronciu, H.J. Wünsche, J. Sieber, K. Schneider, F. Henneberger, Dynamics of single mode semiconductor lasers with passive dispersive reflectors. Opt. Commun. **182**, 221 (2000)
37. J. Guckenheimer, P. Holmes, Nonlinear Oscillations, Dynamical Systems, and Bifurcations of Vector Fields. (Springer, Berlin, 1986)

Chapter 8
Stabilization of Anti-Phase Orbits

Following [1], we study in this section diffusively coupled Hopf normal form oscillators. By introducing a noninvasive delay coupling we are able to stabilize the inherently unstable anti-phase orbits. For the super- and subcritical case we state a condition on the oscillator's nonlinearity, which is necessary and sufficient to find coupling parameters for successful stabilization. We prove these conditions and review previous results on the stabilization of odd-number orbits by time-delayed feedback. Finally, we illustrate the results with numerical simulations.

8.1 Two Diffusively Coupled Oscillators

The model we want to study is given by

$$\dot{z}_1 = f(z_1) + a \cdot (z_2 - z_1), \tag{8.1a}$$

$$\dot{z}_2 = f(z_2) + a \cdot (z_1 - z_2). \tag{8.1b}$$

Here the $z_1, z_2 \in \mathbb{R}^2 \cong \mathbb{C}$ describe the state of each oscillator, $a > 0$ is the diffusive coupling constant, and

$$f(z) = (\lambda + i + \Gamma |z|^2)z \tag{8.2}$$

is the normal form of a Hopf bifurcation as studied in Chap. 3. Note, however that in this case we allow the parameter Γ, describing the nonlinearity of the oscillators, to take on any value in $\mathbb{C}$ in contrast to Chap. 3, where we considered for

Most of this chapter was previously published as B. Fiedler, V. Flunkert, P. Hövel, and E. Schöll, *Delay stabilization of periodic orbits in coupled oscillator systems*, Phil. Trans. R. Soc. A **368**, 1911, 319–341 (2010), reproduced with permission.

simplicity $\Gamma = 1 + i\gamma$, i.e., parameters with unity real part. We denote in the following $\Gamma_i = \text{Im}\Gamma$ and $\Gamma_r = \text{Re}\Gamma$.

For $a = 0$ each oscillator z_j undergoes Hopf bifurcation

$$z_1(t) = z_2(t) = z_+(t) = r_+\exp(2\pi it/p_+) \tag{8.3}$$

with amplitude $r_+^2 = -\lambda/\Gamma_r$ for $\lambda\,\Gamma_r < 0$, as λ increases through the bifurcation point $\lambda = 0$. The Hopf bifurcation is subcritical for fixed $\Gamma_r > 0$, and supercritical for $\Gamma_r < 0$. The period p_+ depends on the amplitude r_+ via

$$\frac{2\pi}{p_+} = 1 + r_+^2\Gamma_i. \tag{8.4}$$

The symmetry $z_1 \longleftrightarrow z_2$ of (8.1) implies that the synchronization manifold is invariant. Let

$$z_\pm = \frac{1}{2}(z_1 \pm z_2) \tag{8.5}$$

denote the symmetrized (+) and the anti-symmetrized (−) variables of the two oscillators. Then in these new variables (8.1) are given by

$$\dot{z}_+ = \frac{1}{2}[f(z_+ + z_-) + f(z_+ - z_-)], \tag{8.6a}$$

$$\dot{z}_- = \frac{1}{2}[f(z_+ + z_-) - f(z_+ - z_-)] - 2az_-. \tag{8.6b}$$

For $z_- = 0$ we have $\dot{z}_- = 0$ and thus the synchronization manifold

$$Z_+ := \{(z_+, z_-)|z_- = 0\}, \tag{8.7}$$

on which $z_1 \equiv z_2$, is invariant. In this context we will call this the *in-phase manifold*, since we only consider oscillations. Because $f(-z) = -f(z)$ is an odd nonlinearity, there is another invariant manifold

$$Z_- = \{(z_+, z_-)|z_+ = 0\}, \tag{8.8}$$

where $z_1 \equiv -z_2$, which we call the *anti-phase manifold*.

For the in-phase dynamics on Z_+ the coupling term vanishes and the dynamics is thus determined by (8.2) and features Hopf bifurcation of the PO $z_+(t)$ with period p_+ as in (8.3) and (8.4). For the anti-phase dynamics, on the other hand, the coupling term does not vanish and the dynamics of z_- is (for $z_+ = 0$ given by)

$$\dot{z}_- = f(z_-) - 2az_-. \tag{8.9}$$

Therefore the anti-phase Hopf bifurcation occurs at $\lambda = 2a$ and generates POs

$$z_-(t) = r_-\exp(2\pi it/p_-) \tag{8.10}$$

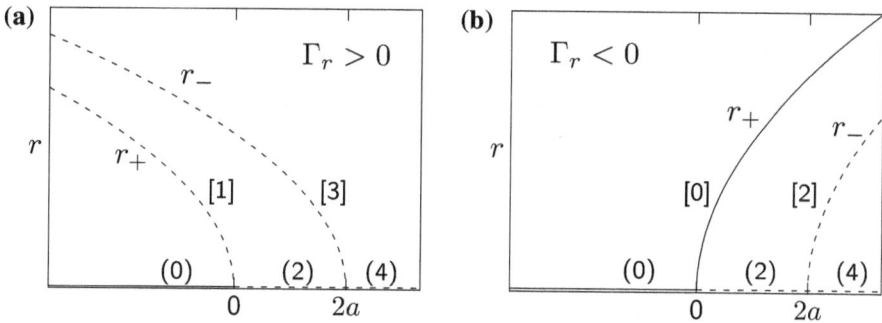

Fig. 8.1 In-phase (r_+) and anti-phase (r_-) Hopf bifurcations $z_\pm(t)$. Dashed and solid curves correspond to unstable and stable solutions, respectively. Panel (**a**) shows the subcritical case and panel (**b**) shows the supercritical case. The unstable dimensions are indicated in parentheses, for the FP $z \equiv 0$, and in brackets for the POs

with amplitude $r_-^2 = -(\lambda - 2a)/\Gamma_r$ and period p_- given by

$$\frac{2\pi}{p_-} = 1 + r_-^2 \Gamma_i. \tag{8.11}$$

Compared to the in-phase dynamics on Z_+ the anti-phase dynamics on Z_- is the same only with a bifurcation parameter shifted by $2a$.

These above observations have an important consequence for the stability properties of the bifurcating anti-phase POs $z_-(t)$. Since $\lambda = 2a > 0$, at the Hopf bifurcation of the anti-phase orbit, the unstable dimension of $z_-(t)$ is at least 2, as inherited from the Hopf bifurcation point at $\lambda = 0$ itself. In particular the unstable dimension is 3 in the subcritical case $\Gamma_r > 0$, and 2 in the supercritical case $\Gamma_r < 0$. The bifurcations are depicted in Fig. 8.1 for the subcritical case (panel (a)) and the supercritical case (panel (b)). As usual the unstable dimension denotes the number of Floquet multipliers strictly outside the complex unit circle and the number in parenthesis denote the unstable dimension of the FP $z = 0$ (cp Chaps. 3 and 4). See [2] and [3, 4, 5] for the mathematical center manifold theory concerning the exchange of stability at bifurcations.

8.2 Stabilization by Delay—Theorems

We aim to stabilize the unstable anti-phase PO $z_-(t)$ by a delayed control term, which is adapted to the specific symmetry $z_+ \equiv 0$ of $z_-(t)$. On the anti-phase PO we have $z_-(t - p_-/2) = -z_-(t)$ (see (8.10)). Therefore $z_+ \equiv 0$ implies the following symmetry with respect to exchanging the two subsystems

$$z_1(t) = z_2\left(t - \frac{p_-}{2}\right), \tag{8.12a}$$

$$z_2(t) = z_1\left(t - \frac{p_-}{2}\right). \tag{8.12b}$$

Indeed, the oscillators switch their roles after half a period in the anti-phase case. This in turn motivates us to seek a stabilization of the solution $z_-(t)$ in the form of delayed coupling

$$\dot{z}_1 = f(z_1) + a \cdot (z_2 - z_1) + b \cdot (z_2(t - \tau) - z_1) \tag{8.13a}$$

$$\dot{z}_2 = f(t_2) + a \cdot (z_1 - z_2) + b \cdot (z_1(t - \tau) - z_2) \tag{8.13b}$$

with a complex coupling strengths $b \in \mathbb{C}$. Note that the delay τ is noninvasive for

$$\tau = \frac{1}{2} n p_-, \tag{8.14}$$

i.e., for integer multiples n of half the period p_-. This delayed coupling is invasive on in-phase solutions z_+, because $z_2(t - \tau) - z_1(t) = z_1(t - \tau) - z_2(t) \neq 0$ there, for half period delays τ, unless $p_+ = p_-/2$.

As the results for stabilization are rather complex we will formulate them in two theorems and give the proofs later in Sect. 8.5.

Theorem 8.1 *Consider the coupled oscillator system* (8.1) *and* (8.2) *with diffusive coupling constant*

$$0 < a < \frac{1}{\pi} \tag{8.15}$$

in the supercritical case $\Gamma_r < 0$. *Then there exists a strictly decreasing real analytic function* $b_* = b_*(a) > a$ *with limits* $b_*(0) = \infty$ *and* $b_*(1/\pi) = 1/\pi$ *such that for real controls*

$$a < b < b_*(a) \tag{8.16}$$

the anti-phase POs $z_-(t) \not\equiv 0, z_+ \equiv 0$ *of* (8.10) *and* (8.11) *are stabilized noninvasively by a delayed coupling* (8.13) *with half period delay*

$$\tau = \frac{1}{2} p_-, \tag{8.17}$$

for small amplitudes $r_- = |z_-(t)|$, *and for parameters* λ *near the anti-phase Hopf bifurcation at* $\lambda = 2a$.

Theorem 8.2 *Consider the subcritical case* $\Gamma_r > 0$ *of Theorem 8.1, again for* $0 < a < 1/\pi$. *Then there exists a continuous, strictly increasing function* $\beta = \beta(a)$ *with limits* $\beta(0) > 0$ *and* $\beta(1/\pi) = \infty$ *such that the following holds for*

$$|\Gamma_i| > \beta(a)\Gamma_r. \tag{8.18}$$

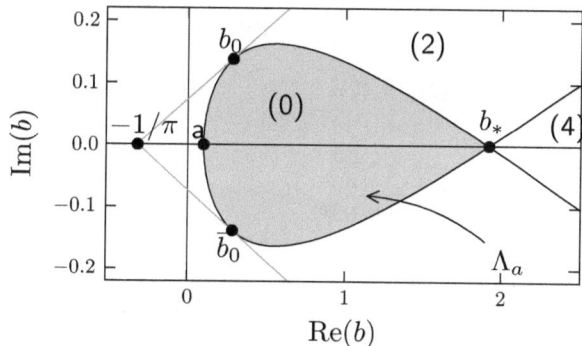

Fig. 8.2 Stabilization region (*shaded*) of complex control coefficients $b \in \mathbb{C} \setminus \mathbb{R}$ for anti-phase solutions near supercritical Hopf bifurcation. Numbers in parentheses indicate the total multiplicity $E(b)$ of eigenvalues with strictly positive real part, at Hopf bifurcation $\lambda = 2a$. Note that straight lines through $b = -1/\pi \in \mathbb{C}$ touch the boundary of the stabilization region in the points b_0 and $\bar{b}_0$ (see Sect. 8.5). Parameter: $a = 0.1$

There exists an open region of controls $b \in \mathbb{C}\backslash\mathbb{R}$, depending on a and Γ, for which (8.13) achieve noninvasive delayed feedback stabilization locally near Hopf bifurcation at $\lambda = 2a$, as asserted in Theorem (8.1).

See Fig. 8.2 for a sketch of the stabilization regime $b \in \mathbb{C} \setminus \mathbb{R}$ to which Theorem 8.2 applies. The shaded region Λ_a indicates the region of those strictly complex controls b, for which stabilization is possible, locally near anti-phase Hopf bifurcation, provided that $|\Gamma_i|/|\Gamma_r|$ is large enough. For decreasing values $|\Gamma_i|/|\Gamma_r| \searrow \beta(a)$ the regime of stabilizing b shrinks to two complex conjugate boundary points $b_0(a), \overline{b_0(a)}$. See Sect. 8.4 for details on $b_*(a)$ and the end of Sect. 8.5 for $b_0(a), \beta(a)$. See Sect. 8.6 for numerical examples.

8.3 Beyond Odd-Number Limitation for Planar Hopf Bifurcation

To prepare for the proof of Theorems 8.1 and 8.2 in Sect. 8.5 we revisit the counterexample

$$\dot{z} = (\lambda + i + \Gamma|z|^2)z + b \cdot (z(t - \tau) - z) \tag{8.19}$$

as discussed in Chap. 3. This time, however, we will apply a slightly different argument utilizing complex analytic maps.

Of course we keep in mind that (8.19) also describes stabilization within the invariant subspace $Z_+ = \{(z_1, z_2) | z_- = 0\}$ of in-phase solutions $z_1(t) \equiv z_2(t)$, introduced in (8.5) and (8.7), under the naive delayed feedback control scheme

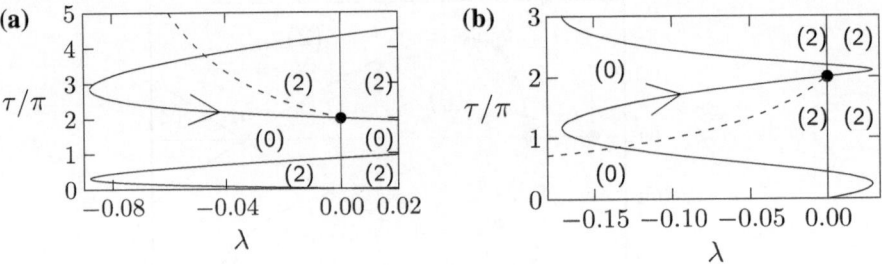

Fig. 8.3 Subcritical Hopf bifurcation in the parameter plane (λ, τ) with fixed control b;**(a)** for soft springs $\Gamma_i < 0$ and **(b)** for hard springs $\Gamma_i > 0$. Solid black lines display the Hopf bifurcation curve $(\lambda(\omega), \tau(\omega))$ emanating from $\lambda = 0$ and noninvasive delay $\tau = 2\pi$. Hopf curves are oriented with increasing ω. The dashed black lines correspond to the period $\tau = p(\lambda)$ of the Pyragas curve of bifurcating periodic solutions. Strict unstable dimensions $E(b)$ of the FP $z \equiv 0$ are indicated in parentheses. Parameters: a) (soft spring): $\Gamma_r = 1, \Gamma_i = -10, b = 0.3e^{i\pi/4}$; b) (hard spring): $\Gamma_r = 1, \Gamma_i = 10, b = 0.1e^{-i3\pi/4}$

$$\dot{z}_1 = f(z_1) + a(z_2 - z_1) + b \cdot (z_1(t - \tau) - z_1) \tag{8.20a}$$

$$\dot{z}_2 = f(z_2) + a(z_1 - z_2) + b \cdot (z_2(t - \tau) - z_2). \tag{8.20b}$$

This case is discussed in Sect. 8.7

Theorem 8.3 *Consider the planar Hopf normal form system* (8.19) *with subcritical Hopf bifurcation at absent control* $b = 0$, *i.e., with*

$$\Gamma_r > 0. \tag{8.21}$$

Then there exist complex control gains b *such that the bifurcating POs* $z(t) = r \exp(2\pi it/p), r^2 = -\lambda/\Gamma_r, p = 2\pi/(1 - \lambda\Gamma_i/\Gamma_r)$ *are stabilized noninvasively by a delayed feedback* (8.19) *with delay equal to the period*

$$\tau = p. \tag{8.22}$$

This holds for small amplitudes r *and for parameters* λ *near Hopf bifurcation at* $\lambda = 0$.

To prepare for our proof of Theorems 8.1 and 8.2 in Sects. 8.2 and 8.3 we now sketch a proof of Theorem 8.3, in the same spirit. For brevity we only consider the *hard spring case*

$$\Gamma_i > 0, \tag{8.23}$$

where period $p = 2\pi/(1 + r^2\Gamma_i)$ decreases with amplitude r. The proof for the *soft spring case* $\Gamma_i < 0$ is very similar.

The basic idea of the proof is easily sketched in the two parameter diagram of Fig. 8.3. There are two ingredients. *First* we linearize at the FP $z \equiv 0$ and study the strict unstable dimensions. Let $E(b)$ denote the total number of eigenvalues η with $\operatorname{Re}\eta > 0$, counting real multiplicities. Even for fixed nonzero b, these numbers still depend on (λ, τ) as indicated in Fig. 8.3 in parentheses. *Second* we evaluate the dashed period curve

$$\tau = p = p(\lambda) = 2\pi/(1 - \lambda\Gamma_i/\Gamma_r) \qquad (8.24)$$

of noninvasive control in Fig. 8.3, as it emanates from the Hopf point $\lambda = 0$, $\tau = 2\pi$ to the subcritical side $\lambda < 0$.

With these two ingredients the proof works as follows. Suppose we can choose b such that the dashed period curve enters a region with $E(b) = 2$ at $\lambda = 0, \tau = 2\pi$, transversely to the Hopf curve and pointing away from the $E(b) = 0$ region. Then the *subcritical* Hopf bifurcation along the λ-axis $\tau = 0$ (alias $b = 0$) has become *supercritical* for chosen parameters along the dashed curve. Hence the bifurcating unstable orbits, for $\tau = 0$, alias $b = 0$, have become stable along the dashed curve of noninvasive delayed feedback control, by standard exchange of stability at Hopf bifurcation.

Let us implement the above idea for our specific case (8.19). Linearization at $z \equiv 0$ yields the characteristic equation

$$0 = \chi(\eta) = \lambda + i + b(e^{-\tau\eta} - 1) - \eta \qquad (8.25)$$

for the eigenvalues η.

Consider the starting point $\lambda = 0, \tau = 2\pi$ with purely imaginary eigenvalue $\eta = i$, first. We determine the strict unstable dimension $E(b)$ there, aiming for $E(b) = 0$ to ensure that the Hopf eigenvalue $\eta = i$ actually effects a change from $E = 0$ to $E = 2$. Any $E > 0$ there would indeed be inherited as an instability of any bifurcating PO, obstructing stabilization. Note here that the Hopf eigenvalue pair $\eta = \pm i$ itself does not yet contribute to the *strict* unstable dimension $E(b)$ at $\lambda = 0, \tau = 2\pi$.

At $\lambda = 0, \tau = 2\pi$ the characteristic equation for $\eta = \tilde{\omega}i =: (1 + \omega)i$ reads

$$b = b(\omega) = \frac{i\omega}{e^{-2\pi i\omega} - 1} = -\frac{\omega}{2}(\cot(\pi\omega) + i). \qquad (8.26)$$

Note $b(0) = -1/(2\pi)$ and the singularities of $b(\omega)$ at integer $\omega \in \mathbb{Z}\setminus\{0\}$. In particular $\eta = 0$ is never an eigenvalue at $\lambda = 0, \tau = 2\pi$, and $E = E(b)$ can only change by Hopf bifurcation there.

Since $E(0) = 0$ at $b = 0$, by planarity and the eliminated trivial Hopf eigenvalue $\eta = i$, the strict unstable dimensions $E(b)$ are as indicated in Fig. 8.4. Indeed complex analytic maps, like $\omega \mapsto b(\omega)$, preserve orientation. Instability $E(b)$ tracks eigenvalues $\operatorname{Re}\eta > 0$ to the right of the imaginary axis $\eta = i\tilde{\omega} = i(1 + \omega)$. Therefore $E(b)$ is larger by two on the right side of any of the oriented curves $\omega \mapsto b(\omega)$, when compared to the left side. Elsewhere $E(b)$ does not change. Therefore all unstable dimensions $E(b)$ in Fig. 8.4 follow from $E(0) = 0$ and the indicated orientations of the solid Hopf curves $\omega \mapsto b(\omega)$ for real ω.

Fig. 8.4 Hopf bifurcation
curves $b = b(\omega)$ (solid) and,
in parentheses, strict unstable
dimensions $E(b)$ at
$\lambda = 0, \tau = 2\pi$. Orientation
arrows indicate increasing ω.
Note the shaded region where
$E(b) = 0$. The '+' and '×'
mark the values of b for
Fig. 8.3 *left* and *right*,
respectively

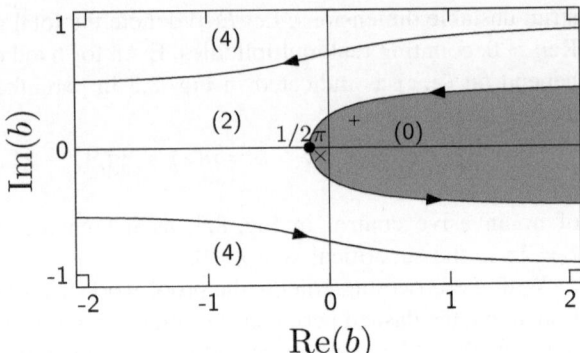

We calculate the tangent $\lambda = \hat{\lambda}, \tau = 2\pi + \hat{\tau}$ to the dashed periodics curve
$\tau = p(\lambda)$ in Fig. 8.3 next. Expanding the explicit representation (8.24) we
immediately obtain

$$\hat{\tau} = 2\pi \frac{\Gamma_i}{\Gamma_r} \hat{\lambda}. \tag{8.27}$$

To compute the tangent $\lambda = \tilde{\lambda}, \tau = 2\pi + \tilde{\tau}$ to the solid Hopf curve $\lambda = \lambda(\omega)$, $\tau =
\tau(\omega)$ of solutions $\eta = i\tilde{\omega} = i(1 + \omega)$ at $\omega = 0$ we linearize the characteristic
equation (8.25), for fixed b, keeping in mind that $\tilde{\lambda}$ and $\tilde{\tau}$ are of order ω. Thus, we
obtain $0 = \tilde{\lambda} + b(-2\pi i\omega - i\tilde{\tau}) - i\omega$ and the tangent of the Hopf curve:

$$\tilde{\lambda} = \frac{\text{Im}b}{\text{Re}b} \omega \tag{8.28a}$$

$$\tilde{\tau} = -\frac{1 + 2\pi \text{Re}b}{\text{Re}b} \omega. \tag{8.28b}$$

To achieve the geometric hard spring situation of Fig. 8.3 right, and hence
prove Theorem 8.3 in the hard spring case, we recall Γ_r and Γ_i are both positive;
see (8.21) and (8.23). Hence (8.27) makes the slope of the dashed periodics curve
positive, in Fig. 8.3. By (8.28) the slope of the solid Hopf curve, in contrast, is
given by

$$\tilde{\tau} = -\frac{1 + 2\pi \text{Re}b}{\text{Im}b} \tilde{\lambda}. \tag{8.29}$$

We now determine the strict unstable dimensions $E(b)$ resulting on different
sides of the Hopf curve $\eta = i\tilde{\omega}$ of the characteristic equation (8.25), for fixed b and
in the (λ, τ)-plane. It is advisable to proceed with analytic care here. Let $(\check{\lambda}, \check{\tau}, \check{\eta}) \in
\mathbb{R}^2 \times \mathbb{C}$ denote infinitesimal variations of (λ, τ, η). By multivariate linearization of
(8.25) at (λ, τ, η) we obtain the equivalent system

$$\varphi(\check{\lambda}, \check{\tau}) := \check{\lambda} - \eta b e^{-\tau\eta} \check{\tau} = \check{\zeta} \tag{8.30a}$$

$$\psi(\breve{\eta}) := (1 + \tau b e^{-\tau \eta})\breve{\eta} = \breve{\zeta} \tag{8.30b}$$

with real linear maps $\varphi : \mathbb{R}^2 \to \mathbb{C}$ and $\psi : \mathbb{C} \to \mathbb{C}$. In other words,

$$\breve{\eta} \mapsto (\breve{\lambda}, \breve{\tau}) = (\varphi^{-1} \circ \psi)(\breve{\eta}), \tag{8.31}$$

if we eliminate the dummy variable $\breve{\zeta} \in \mathbb{C}$.

The map ψ preserves real orientation, being just a multiplication by $1 + \tau b e^{-\tau \eta} \in \mathbb{C}$, for nonzero $1 + \tau b e^{-\tau \eta}$. The map φ, however, when viewed as a linear map $\varphi : \mathbb{R}^2 \to \mathbb{C} \cong \mathbb{R}^2$ by $\mathrm{Re}\varphi$ and $\mathrm{Im}\varphi$, possesses determinant

$$
\begin{aligned}
\det\varphi &= -\mathrm{Im}(\eta b e^{-\eta\tau}) \\
&= -\tilde{\omega}(\mathrm{Re}(b)\cos(\tilde{\omega}\tau) + \mathrm{Im}(b)\sin(\tilde{\omega}\tau)) \\
&= -\mathrm{Re}b,
\end{aligned} \tag{8.32}
$$

at the point of interest $\eta = i\tilde{\omega}, \tilde{\omega} = 1, \lambda = 0, \tau = 2\pi$. In particular φ preserves orientation for $\mathrm{Re}b < 0$ as chosen in Fig. 8.3 right. Therefore $\varphi^{-1} \circ \psi$ also preserves orientation in this case.

By (8.28) and $\mathrm{Im}b < 0$, the Hopf curve is oriented to the upper right at $\lambda = 0$, $\tau = 2\pi$, as indicated in Fig. 8.3, right. Because $\varphi^{-1} \circ \psi$ from (8.31) preserves orientation, $E(b) = 2$ again holds to the right of the oriented Hopf curve $\omega \mapsto (\lambda(\omega), \tau(\omega))$, and the strict unstable dimensions of Fig. 8.3, right, follow. The stabilizing condition for the dashed curve $\tau = p(\lambda)$ of the Pyragas PO [3] to enter the $E(b) = 2$ region, when emanating from $\lambda = 0, \tau = 2\pi$ to the lower left therefore reads

$$2\pi \frac{\Gamma_i}{\Gamma_r} = \frac{\hat{\tau}}{\hat{\lambda}} > \frac{\tilde{\tau}}{\tilde{\lambda}} = -\frac{1 + 2\pi \, \mathrm{Re}b}{\mathrm{Im}b}. \tag{8.33}$$

Clearly this can always be achieved in any subcritical, soft spring case of positive Γ_r and Γ_i, if we choose $1 + 2\pi\mathrm{Re}b > 0$ small enough with $\mathrm{Re}b \gtrsim -1/2\pi$, and $\mathrm{Im}b < 0$ also negative, such that b resides in the lower left part of the shaded region indicated in Fig. 8.4.

We thus have achieved supercritical Hopf bifurcation along the dashed line of noninvasive delayed feedback in Fig. 8.3, and hence local stability of the bifurcating branch. This proves Theorem 8.3.

8.4 Characteristic Equations for Two Coupled Oscillators

In this section we return to our control system (8.13) of coupled oscillators, linearized at $z_1 \equiv z_2 \equiv 0$. In terms of the coordinates $z_\pm = (z_1 \pm z_2)/2$ from 8.5) the linearization reads

$$\dot{z}_+ = (\lambda + i)z_+ + b \cdot (z_+(t - \tau) - z_+) \tag{8.34a}$$

$$\dot{z}_- = (\lambda - 2a + i)z_- - b \cdot (z_-(t - \tau) + z_-). \tag{8.34b}$$

Note how the linearization decouples into $Z_\pm$-components $z_\pm$, just as in the case $b = 0$ of absent control; see (8.5)–(8.11). In the previous section noninvasive delay stabilization of local subcritical Hopf bifurcation was achieved once that Hopf point itself was stabilized. Analogously we now study stability at the Hopf point $\lambda = 2a$ itself, before addressing the bifurcating anti-phase orbits in the following section.

The exponential ansatz $z_\pm = \exp(\eta t)$ yields the following two characteristic equations:

$$0 = \chi_+(\eta) = \lambda + i + b(e^{-\tau\eta} - 1) - \eta \tag{8.35a}$$

$$0 = \chi_-(\eta) = \lambda - 2a + i - b(e^{-\tau\eta} + 1) - \eta. \tag{8.35b}$$

Because the linearization (8.34) decouple in $z_\pm$, each of (8.35) contributes its own independent set of eigenvalues to the total spectrum; the strict unstable dimensions $Re\eta > 0$ of χ_+ and χ_- therefore add up to the total unstable dimension $E(b) = E_+(b) + E_-(b)$ of the trivial FP, see (8.4) below.

For $b = 0$ we find a Hopf bifurcation in $Z_- = \{(z_+, z_-) | z_+ = 0\}$ at $\lambda = 2a$ and $\eta = i$, i.e., for period $p_- = 2\pi$. Therefore $\tau = p_-/2 = \pi$ at the Hopf bifurcation, and (8.35) become

$$0 = \chi_+(\eta) = 2a + i + b(e^{-\pi\eta} - 1) - \eta \tag{8.36a}$$

$$0 = \chi_-(\eta) = i - b(e^{-\pi\eta} + 1) - \eta. \tag{8.36b}$$

Consider $b = 0$ first. The complex notation which we have employed then provides a single eigenvalue $\eta = 2a + i$ with positive real part in $Z_+ = \{(z_+, z_-) | z_- = 0\}$ from (8.36a). In Z_- the characteristic equation (8.36b) provides the simple Hopf eigenvalue $\eta = i$, expectedly. Let $E(b)$ again denote the total number of eigenvalues η with $Re\eta > 0$, adding both Z_+ and Z_- and counting real multiplicities. Then we have just proved

$$E(0) = 2 \tag{8.37}$$

at $\lambda = 2a, \tau = \pi$, for this *strict* unstable (or expanding) dimension $E(b)$.

Could the unstable dimension $E(b)$ change, as b varies? To achieve our goal

$$E(b) = 0 \tag{8.38}$$

of Hopf stabilization at $\lambda = 2a, \tau = \pi$, it better changes, somehow.

We first note $E(b) = 2$ for small $|b|$. Indeed the delay exponential $\exp(-\pi\eta)$ then just generates a plethora of countably infinitely many discrete eigenvalues η, in each of the characteristic equations (8.36a) and (8.36b), all with large negative real part.

Can $E(b)$ change by an eigenvalue η crossing zero as b varies? Inserting $\eta = 0$ in (8.36a) and (8.36b) shows that this cannot happen via χ_+. In χ_- however, $\eta = 0$ is a solution if and only if

$$b = \frac{i}{2}. \tag{8.39}$$

It remains to study changes of $E(b)$ by a purely imaginary Hopf eigenvalue

$$\eta = i\tilde{\omega}. \tag{8.40}$$

Let $E_{\pm}(b)$ count the solutions η with $\mathrm{Re}\,\eta > 0$ of $\chi_{\pm}(\eta) = 0$, with algebraic multiplicity, so that

$$E(b) = E_{+}(b) + E_{-}(b). \tag{8.41}$$

Since $E_{\pm} \geq 0$ and $E_{+}(0) = 2, E_{-}(0) = 0$ we study changes of $E_{+}(b)$ via $\eta = i\tilde{\omega}$, first, hoping for a region of $b \in \mathbb{C}$ where $E(b) = 0$. Solving (8.36a) with

$$\eta = i\tilde{\omega} = i(1 + 2\omega) \tag{8.42}$$

we obtain the Hopf curves

$$b = b_{+}(\omega) = 2\frac{a - i\omega}{1 + \exp(-2\pi i\omega)} \tag{8.43}$$
$$= a + \omega\tan(\pi\omega) + i(-\omega + a\tan(\pi\omega))$$

with singularities at odd integers 2ω. See the solid lines of Fig. 8.5 for a sketch of these Hopf curves, when $a = 0.1$.

Solving (8.36b) with $\eta = i\tilde{\omega} = i(1 + 2\omega)$ we obtain the Hopf curves

$$b = b_{-}(\omega) = 2\frac{i\omega}{\exp(-2\pi i\omega) - 1} = -\omega(\cot(\pi\omega) + i) \tag{8.44}$$

with singularities at integer $\omega \neq 0$. See the dashed lines of Fig. 8.5. Note how these dashed lines correspond to the solid lines of Fig. 8.4 because (8.44) corresponds to (8.26) in the sense that $b_{-}(\omega) = 2b(\omega)$.

To determine the changes of the real unstable dimensions $E(b) = E_{+}(b) + E_{-}(b)$ along the curves $\omega \mapsto b_{\pm}(\omega)$ we observe that the zeros of $\chi_{\pm}$ are given as complex analytic functions. An elementary calculation shows that the complex derivative $b'_{-}(\omega)$ never vanishes. The complex derivative $b'_{+}(\omega)$ vanishes if, and only, if

$$a = \frac{1}{\pi} \text{ and } \omega = 0. \tag{8.45}$$

This is precisely where the shaded loop Λ_a of Fig. 8.5 is formed. For

$$0 < a < \frac{1}{\pi} \tag{8.46}$$

this loop stretches over the real interval

$$a < b < b_{*}(a), \tag{8.47}$$

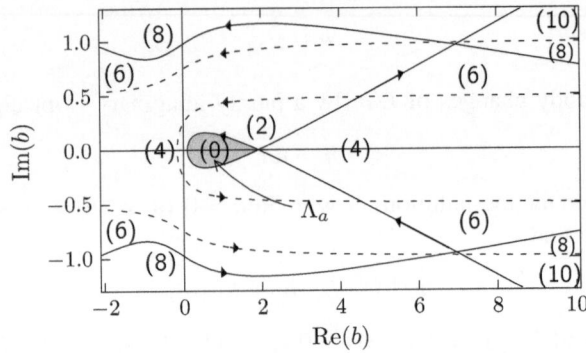

Fig. 8.5 Oriented Hopf curves $b = b_+(\omega)$, solid, and $b = b_-(\omega)$, dashed, for $\omega \in \mathbb{R}$ in control system (8.13), linearized at $z_1 \equiv z_2 \equiv 0$. Orientation arrows indicate increasing ω. Note symmetry with respect to the real axis, due to complex conjugation $b_\pm(-\omega) = \overline{b_\pm(\omega)}$. Unstable dimensions are indicated by $(E(b))$. Note $E(b) = 0$ inside the shaded loop Λ_a. Parameter: $a = 0.1$

where $b_*(a) = b_+(\omega_*(a)) > 0$ is given by (8.43) evaluated at the first positive solution $\omega_* = \omega_*(a) > 0$ of the transcendental equation

$$0 = \mathrm{Im}(b_+(\omega_*)) = -\omega_* + a\tan(\pi\omega_*). \tag{8.48}$$

To determine $E(b)$ in Fig. 8.5 we proceed as for Fig. 8.4. Complex analytic maps like $\omega \mapsto b_\pm(\omega)$, with nonvanishing derivatives, preserve orientation. Instability $E(b) = E_+(b) + E_-(b)$ tracks eigenvalues $\mathrm{Re}\,\eta > 0$, i.e., η to the right of the imaginary axis $\eta = i\tilde{\omega} = i(1 + 2\omega)$. (The only zero eigenvalue $\eta = 0$ at $\omega = -1/2$ of $b_-(-1/2) = i/2$ noted in (8.39) makes no exception here). Therefore $E(b)$ is larger by two on the right side of any of the oriented curves $\omega \mapsto b = b_\pm(\omega)$, when compared to the left side. Elsewhere $E(b)$ does not change. Starting from $E(0) = 2$, as noted in (8.37), it is therefore elementary to derive all strict real unstable dimensions $E(b)$ of the Hopf bifurcation point $\lambda = 2a$, as given in Fig. 8.5.

In particular $E(b) = 0$ if and only if b is inside the shaded loop Λ_a of Fig. 8.5, and that stabilizing loop exists if and only if $0 < a < 1/\pi$.

8.5 Proof of Stabilization Theorems

Based on the analysis of the strict unstable dimension $E(b)$ at the anti-phase Hopf bifurcation $\lambda = 2a, \tau = \pi, \eta = i$ as given in the previous section, we now proceed to prove local noninvasive delayed feedback stabilization of the bifurcating anti-phase periodic solutions, as claimed in Theorems 8.2 and 8.2 for the supercritical and the subcritical case, respectively. Both proofs are based on the strategy of Sect. 8.3 The stabilization region $E(b) = 0$ of Fig. 8.4 for the complex control $b \in \mathbb{C}$ now has to be replaced by the shaded loop region Λ_a of $E(b) = 0$ derived in

Fig. 8.2 and, in further detail, in Fig. 8.5. We recall that the loop Λ_a is bounded by the section $|\omega| \leq \omega_*$ of the curve

$$b_+(\omega) = a + \omega \tan(\pi\omega) + i(-\omega + a \tan(\pi\omega)), \qquad (8.49)$$

where $\omega_* = \omega_*(a)$ is the first positive solution of

$$0 = \omega_* - a \tan(\pi\omega_*). \qquad (8.50)$$

See (8.43) and (8.48). We denote $b_*(a) := b_+(\omega_*(a)) > 0$, as above.

Analogously to Sect. 8.3, Fig. 8.3, we now seek the geometric situation of Fig. 8.6 for the tangents and unstable dimensions of the solid oriented Hopf curve $(\lambda(\omega), \tau(\omega))$ and the dashed curve $\tau = p_-(\lambda)/2$ of bifurcating periodic solutions. We calculate their tangents at $\lambda = 2a, \tau = \pi, \eta = i$ next; see also (8.27) versus (8.28).

Linearizing the explicit representation

$$\tau = \frac{1}{2}p_-(\lambda) = \frac{\pi}{1 - (\lambda - 2a)\Gamma_i/\Gamma_r} \qquad (8.51)$$

of (8.10), (8.11), and (8.17) at $\lambda = 2a, \tau = \pi$ with $\lambda = 2a + \hat{\lambda}, \tau = \pi + \hat{\tau}$, we obtain

$$\hat{\tau} = \pi \frac{\Gamma_i}{\Gamma_r} \hat{\lambda} \qquad (8.52)$$

for the tangent to the anti-phase periodics, in analogy to (8.27). Analogously to (8.28) we also linearize the characteristic equation (8.35b), for fixed b, and obtain $0 = \chi_-(\eta) = \tilde{\lambda} - b(2\pi i\omega + i\tilde{\tau}) - 2i\omega$ with $\eta = i(1 + 2\omega), \lambda = 2a + \tilde{\lambda}, \tau = \pi + \tilde{\tau}$. This yields the tangent to the anti-phase Hopf curve:

$$\tilde{\lambda} = \frac{\mathrm{Im}b}{\mathrm{Re}b} 2\omega \qquad (8.53a)$$

$$\tilde{\tau} = -\frac{1 + \pi\mathrm{Re}b}{\mathrm{Re}b} 2\omega. \qquad (8.53b)$$

We now address the supercritical case $\Gamma_r < 0$ of theorem 8.1 In Sect. 8.4 we have seen how the stabilizing loop Λ_a arises for $0 < a < 1/\pi$; see assumption (8.15). We claim that all real controls b in Λ_a stabilize the local anti-phase branch of periodic solutions, i.e., all $a < b < b_*(a)$; see (8.16) and the explicit representation of $b_*(a)$ in (8.47) and (8.48) as well as (8.49) and (8.50). Analyticity, strict monotonicity, and the claimed limits of $b_*(a)$ are obvious.

To prove stabilization note that the slope of the anti-phase Hopf curve (8.53) is vertical in the (λ, τ)-plane, for $\mathrm{Im}b = 0$. Moreover the orientation is downwards, decreasing $\tilde{\tau}$, since $\mathrm{Re}b > 0$ in the loop Λ_a. It remains to prove $E = 2$ along any periodics curve emanating to the supercritical right of the τ-axis and stabilization of the bifurcating anti-phase periodic orbits will follow, as claimed in Theorem 8.1.

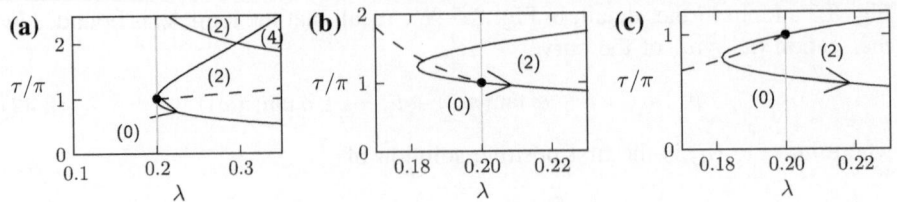

Fig. 8.6 Anti-phase Hopf bifurcation in the parameter plane (λ, τ) for fixed control b. Solid line: Hopf bifurcation curve $(\lambda(\omega), \tau(\omega))$ through $\lambda = 0$ and noninvasive delay $\tau = \pi$. Dashed line $\tau = \frac{1}{2}p_-(\lambda)$: Pyragas curve of bifurcating periodic solutions. Panel (**a**): supercritical case $\Gamma_r < 0$ and $a < b < b_*(a) := b_+(\omega_*(a))$. Panel (**b**): subcritical soft spring case $\Gamma_r > 0, \Gamma_i < 0$. Panel (**c**): subcritical hard spring case $\Gamma_r > 0, \Gamma_i > 0$. Parameters: (a) $\Gamma_r = -1$, $\Gamma_i = -1$, $b = 1.0$ ($b_* \approx 1.92$); (b) $\Gamma_r = +1$, $\Gamma_i = -10$, $b = 0.24e^{+i\pi/8}$; (c) $\Gamma_r = +1$, $\Gamma_i = +10$, $b = 0.24e^{-i\pi/8}$; $a = 0.1$ in all plots

To determine the regions $E = 2$, we proceed as in Sect. 8.3, (8.30)–(8.32) but with modified real linear maps

$$\varphi(\breve{\lambda}, \breve{\tau}) = \breve{\lambda} + \eta b e^{-\tau\eta}\breve{\tau} = \breve{\zeta} \tag{8.54a}$$

$$\psi(\breve{\eta}) = (1 - \tau b e^{-\tau\eta})\breve{\eta} = \breve{\zeta}, \tag{8.54b}$$

which arise from the modified characteristic equation (8.35b) for $\chi_-(\eta) = 0$, replacing (8.25). Again

$$(\breve{\lambda}, \breve{\tau}) = (\varphi^{-1} \circ \psi)(\breve{\eta}), \tag{8.55}$$

where ψ preserves real orientation. At the point of interest $\eta = i\tilde{\omega}$, $\tilde{\omega} = 1$, $\lambda = 0$, and $\tau = \pi$ this time, the linear map $\varphi : \mathbb{R}^2 \to \mathbb{C} \cong \mathbb{R}^2$ again possesses determinant

$$\begin{aligned} \det\varphi &= +\mathrm{Im}(\eta b e^{-\tau\eta}) \\ &= +\tilde{\omega}(\mathrm{Re}(b)\cos(\tilde{\omega}\tau) + \mathrm{Im}(b)\sin(\tilde{\omega}\tau)) \\ &= -\mathrm{Re}b, \end{aligned} \tag{8.56}$$

because $\tilde{\omega}\tau = \pi$, this time. Since $\mathrm{Re}b > 0$ in the loop Λ_a of linear in-phase stabilization of Fig. 8.4, the map φ *reverses* orientation, this time, and so does the composition $\varphi^{-1} \circ \psi$ of (8.55). Therefore the region $E = 2$ now appears *to the left* of the Hopf curve in the (λ, τ)-plane. Since the Hopf curve is oriented vertically downwards, the $E = 2$ region contains the tangent of any supercritical Pyragas curve $\tau = \frac{1}{2}p_-(\lambda)$ emanating to the right of the τ-axis; see Fig. 8.6 left. This proves Theorem 8.1.

Finally we settle the subcritical case $\Gamma_r > 0$ of Theorem 8.2 Fix $0 < a < 1/\pi$ and consider strictly complex b in the stability loop Λ_a of Fig. 8.5, first in the soft spring case

$$\Gamma_i < 0. \tag{8.57}$$

We determine those Γ_i next, for which such choices of b are able to stabilize the local bifurcating anti-phase branch, noninvasively. From the orientation analysis of (8.54)–(8.56) we again conclude $E = 2$ *to the left* of the oriented Hopf curve in the (λ, τ)-plane. Analogously to (8.33) it is therefore immediate that we encounter the stabilizing soft spring geometric situation of Fig. 8.6, center, if and only if the Hopf slope $\tilde{\tau}/\tilde{\lambda}$ of (8.53) exceeds the slope $\hat{\tau}/\hat{\lambda}$ of the periodics (8.52), i.e.

$$0 > -\pi\frac{\text{Re}b + 1/\pi}{\text{Im}b} > \pi\frac{\Gamma_i}{\Gamma_r}. \tag{8.58}$$

In particular $\text{Im}b$ is required to be positive for the proper orientation of the Hopf curve. The least restrictive choice is given by $\Lambda_a \ni b \to b_0(a)$ defined such that the minimum

$$\beta(a) = \min\left\{\frac{\text{Re}b + 1/\pi}{\text{Im}b} \,\middle|\, b \in \Lambda_a, \text{Im}b > 0\right\} \tag{8.59}$$

over the closure of the upper half loop $\Lambda_a \cap \{\text{Im}b > 0\}$ is attained, at complex $b = b_0(a)$. Note that $b_0(a)$ is the tangent point where straight lines through $b = -1/\pi \in \mathbb{C}$ touch the boundary of the upper half of the stabilizing loop Λ_a (see Fig. 8.2). This allows us to stabilize the local anti-phase branch for all subcritical soft spring Γ such that

$$0 < \beta(a)\,\Gamma_r < |\Gamma_i|. \tag{8.60}$$

It is completely analogous to consider stabilization in the subcritical hard spring case $\Gamma_r > 0, 0 < a < 1/\pi$, where

$$\Gamma_i > 0. \tag{8.61}$$

We then arrive at the stabilizing geometric situation of Fig. 8.6, right, for slopes

$$0 < -\pi\frac{\text{Re}b + 1/\pi}{\text{Im}b} < \pi\frac{\Gamma_i}{\Gamma_r} \tag{8.62}$$

and suitable b in the lower half loop $\Lambda_a \cap \{\text{Im}b < 0\}$, again if and only if (8.60) holds. Indeed the symmetry $b_+(-\omega) = \overline{b_+(\omega)}$ of the loop Λ_a implies

$$\beta(a) = \min\left\{\frac{\text{Re}b + 1/\pi}{\text{Im}b} \,\middle|\, b \in \Lambda_a, \text{Im}b > 0\right\} \tag{8.63a}$$

$$= \min\left\{-\frac{\text{Re}b + 1/\pi}{\text{Im}b} \,\middle|\, b \in \Lambda_a, \text{Im}b < 0\right\}. \tag{8.63b}$$

Equation (8.60) therefore allows us to locally stabilize the supercritical anti-phase branch for both the soft and hard spring case, as was claimed in (8.18).

Fig. 8.7 Plotted is $\beta(a)$ (see 8.63). The inset shows the function in a wider range. Note that β is monotonically increasing and $\beta(0) > 0$

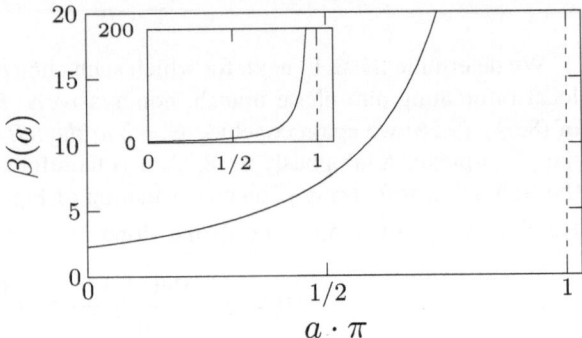

We conclude with deriving the claimed monotonicity and continuity of the minimum function $\beta(a)$ (see Fig. 8.7) It is sufficient to show that the loops Λ_a strictly shrink with increasing a. Then the maximum slopes $1/\beta(a)$ of straight lines through $b = -1/\pi \in \mathbb{C}$ and points of $\Lambda_a \cap \{\mathrm{Im}\, b > 0\}$ likewise decrease.

To show that the loops Λ_a strictly shrink with increasing a we consider the map

$$(\omega, a) \mapsto (\mathrm{Re}\, b_+(\omega), \mathrm{Im}\, b_+(\omega)), \tag{8.64}$$

which defines the boundary of the loop Λ_a for $0 \leq \omega \leq \omega_*(a) < 1/2$; see (8.49) and (8.50). It is easy to check that the map (82) is an orientation preserving local diffeomorphism on $0 \leq a < 1/\pi, |\omega| < 1/2$. With the given counter-clockwise orientation of each loop boundary, by ω, this shows that the loops Λ_a shrink to $b = 1/\pi$ for $a \nearrow 1/\pi$. This completes the proof of Theorems 8.1 and 8.2.

8.6 Numerical Illustrations

In this section we present some numerical results. Figure 8.8 explains the stabilization of the anti-phase orbit for the subcritical case by looking at all rotating waves (circular orbits) present in the system. Plotted are the radii of rotating waves within the anti-phase manifold. The target orbit is stabilized through a transcritical bifurcation with a delay induced rotating wave. For feedback strength a little above the stabilization the FP loses its stability in a subcritical Hopf bifurcation. In the limit $\lambda \to 2a$ the Hopf bifurcation and the transcritical bifurcation occur at the same coupling strengths b_0 resulting in an instant exchange of stability.

Figure 8.9 displays exemplary time series for the stabilized anti-phase circular orbit in the subcritical case. When the target circular orbit is stabilized the control signal vanishes, demonstrating the noninvasiveness of the method.

Figure 8.10 demonstrates how the stabilization fails if condition (8.18) $|\Gamma_i| > \beta(a)\Gamma_r$ is not satisfied. In this case we have chosen $\Gamma = 1 - 4i$. Since $\beta(0.1) \approx 4.37$ the choice of Γ slightly violates the inequality and leads to oscillator deaths instead of stabilized anti-phase orbits.

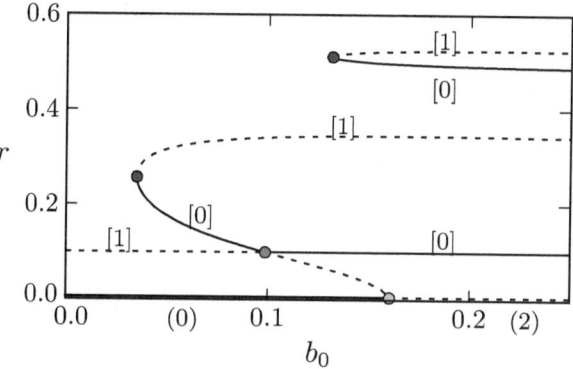

Fig. 8.8 Stabilization of the anti-phase branch in the subcritical case. Plotted are the radii of circular orbits vs. the feedback gain b_0 within the anti-phase manifold; solid and dashed lines correspond to dynamically stable and unstable circular orbits, respectively. For increasing feedback strengths b_0 a pair of stable and unstable orbits is born in a saddle-node bifurcation ($b_0 \approx 0.03$). The stable sibling then stabilizes the target orbit in a transcritical bifurcation (red circle) at ($b_0 \approx 0.1$) and subsequently, having lost its stability, destabilizes the FP $r = 0$ in a subcritical Hopf bifurcation at $b_0 \approx 0.17$ (green circle). Note that the control is noninvasive on the target orbit, i.e., not changing its radius. With further increasing feedback strengths there is a cascade of saddle-node bifurcations (blue circles) generating new feedback-induced circular orbits. One of these bifurcations is shown ($b_0 \approx 0.14$). Unstable dimensions are indicated in parentheses, for the FP $z \equiv 0$, and in brackets, for the bifurcating periodic orbits. Parameters: $a = 0.1, \lambda = 2a - 0.01, \Gamma = 1 - 10i, \tau = p_-/2 = \pi/(1 - (\lambda - 2a)\Gamma_i/\Gamma_r), b = b_0 e^{i\pi/8}$

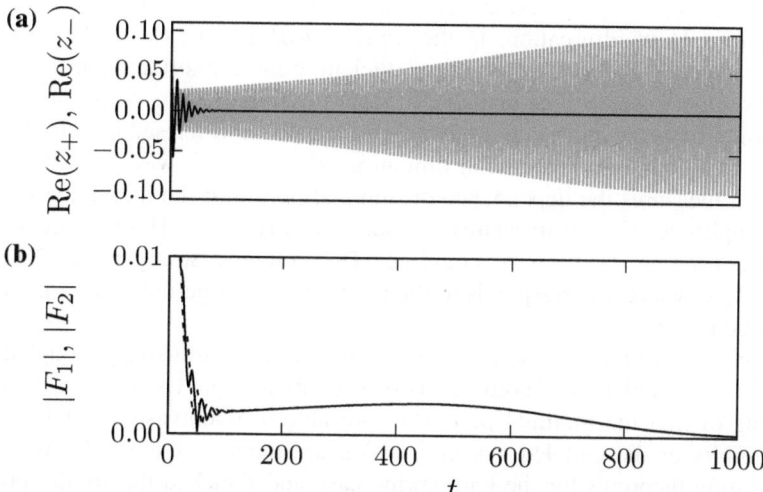

Fig. 8.9 Stabilization of the anti-phase branch in the subcritical case. Panel **(a)**: time series of $\mathrm{Re}z_+$ (*black*) and $\mathrm{Re}z_-$ (*gray*). Panel **(b)**: Time series of coupling forces $F_1 := b \cdot (z_1(t - \tau) + z_2)$ (solid) and $F_2 := b \cdot (z_2(t - \tau) + z_1)$ (*dotted*) acting on the systems. The system starts away from the anti-phase manifold. After a short time the in-phase component decays and the system goes to the anti-phase manifold. After a longer transient the system approaches the stabilized anti-phase orbit. Once the anti-phase orbit is reached the control forces vanishes. Parameters: $a = 0.1, \lambda = 2a - 0.01, \Gamma = 1 - 10i, \tau = p_-/2 = \pi/(1 - (\lambda - 2a)\Gamma_i/\Gamma_r), b = 0.24e^{i\pi/8}$

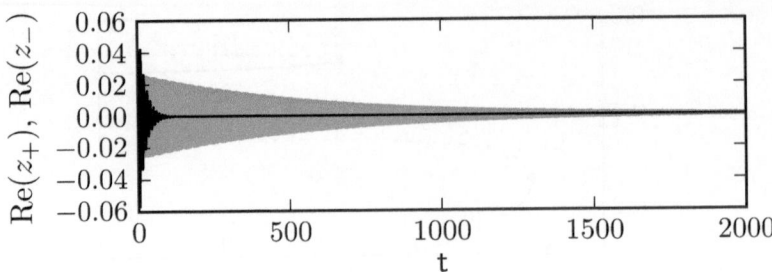

Fig. 8.10 Plotted is the time series of Rez_+ (*black*) and Rez_- (*gray*) for same parameters as Fig. 8.9 except $\Gamma = 1 - 4i$. This Γ violates (8.18) and fails to stabilize the anti-phase circular orbit

8.7 Discussion

The in-phase orbits are stable and unstable for the super- and subcritical case, respectively. Since the dynamics in the in-phase manifold is equivalent to the dynamics of a single system, the stabilization of the in-phase orbit in the subcritical case reduces to the stabilization of an orbit born in a planar subcritical Hopf bifurcation by delayed feedback, which has previously been demonstrated [3]. We are now also able to stabilize unstable anti-phase orbits in certain combinations of super-/subcritical and soft/hard spring cases. Stabilization is achieved locally near Hopf bifurcation. In the *supercritical* case of two unstable Floquet multipliers, real feedback gains were sufficient, both, for soft and hard springs. In the *subcritical* case of three unstable Floquet multipliers, in contrast, stabilization could only be achieved by complex feedback gains and for sufficiently nonlinear (soft or hard) springs. The crucial limitation $|\Gamma_i| > \beta(a)\Gamma_r$ was derived in (8.18). Here $|\Gamma_i|$ measures the dependence of minimal period of the individual oscillator upon amplitude, $\Gamma_r > 0$ measures the subcriticality of the Hopf bifurcation and a is the strength of diffusive coupling. The coupling strength was limited to $0 < a < 1/\pi$, where π corresponds to the normalized half period at anti-phase Hopf bifurcation.

Conclusion—In conclusion we have studied two diffusively coupled Hopf normal form oscillators of both super- or subcritical type. By introducing a delay coupling of half the minimal period we are able to noninvasively stabilize anti-phase orbits of the coupled systems, which are inherently unstable. We proved stabilization theorems for the hard spring case and sketched the similar proof for the soft spring case.

Outlook—Building on this work it will be interesting to apply our method to stabilize anti-phase orbits in physical and biological systems such as coupled lasers and coupled neurons. From a mathematical point of view generalizations to n-oscillators and thus n-fold symmetries may be an interesting direction to pursue [6, 7]. For the Pyragas control scheme of planar in-phase circular orbits it has been shown [8] that only orbits whose real Floquet multipliers μ obey $\mu < \exp(9)$ can be stabilized. Similar Floquet constraints may apply for our control scheme. Although we believe our noninvasive stabilization strategy to be well adapted to anti-phase periodic oscillations, only the derivation and comparison of such fundamental constraints will settle our quest for efficient noninvasive feedback stabilization of spatio-temporal patterns.

References

1. B. Fiedler, V. Flunkert, P. Hövel, E. Schöll, Delay stabilization of periodic orbits in coupled oscillator systems. Phil. Trans. R. Soc. A **368**, 319 (2010)
2. O. Diekmann, S.A. van Gils, S.M. Verduyn Lunel, H.O Walther, *Delay Equations*. (Springer-Verlag, New York, 1995)
3. B. Fiedler, V. Flunkert, M. Georgi, P. Hövel, E. Schöll, Refuting the odd number limitation of time-delayed feedback control. Phys. Rev. Lett. **98**, 114101 (2007)
4. Schöll E. Schuster H. G (eds.)*Handbook of Chaos Control* (Wiley-VCH, Weinheim, 2008), second completely revised and enlarged edition
5. W. Just, B. Fiedler, V. Flunkert, M. Georgi, P. Hövel, E. Schöll, Beyond odd number limitation: a bifurcation analysis of time-delayed feedback control. Phys. Rev. E **76**, 026210 (2007)
6. O. D'Huys, R. Vicente, T. Erneux, J. Danckaert, I. Fischer, Synchronization properties of network motifs: Influence of coupling delay and symmetry. Chaos **18**, 037116 (2008)
7. C.-U. Choe, T. Dahms, P. Hövel, E. Schöll, Controlling synchrony by delay coupling in networks: from in-phase to splay and cluster states. Phys. Rev. E **81**, 025205(R) (2010)
8. B. Fiedler, Time-delayed feedback control: Qualitative promise and quantitative constraints. *Proceedings of the 6th EUROMECH Nonlinear Dynamics Conference*, ENOC-2008, ed. by A. Fradkov, B. Andrievsky, http://lib.physcon.ru/?item=1568

Part II
Synchronization of Delay Coupled Systems

Chapter 9
Introduction

Synchronization phenomena of coupled nonlinear oscillators are omnipresent and play an important role in physical, chemical and biological systems [1–4]. Understanding the synchronization mechanisms is crucial for many practical applications.

One of the most intriguing effects occurring in coupled nonlinear systems is the synchronization of chaotic dynamics [5]. The notions of chaos and synchronization apparently contradict each other. A system is chaotic if small perturbations of the systems initial condition are amplified resulting in an unpredictable dynamical behavior (see Sect. 15.3). Stable synchronization of two systems on the other hand occurs when deviations between the system states decay with time (negative transversal Lyapunov exponent).

The contradiction between these two characteristics is only apparent because the decay and amplification occur in different directions in phase space. Perturbations within the synchronization manifold (SM), i. e., the manifold on which the states of the two systems are identical, grow due to a positive Lyapunov exponent (LE) within this manifold giving rise to the chaotic dynamics. On the other hand, perturbations orthogonal to the SM, associated with deviations between the two systems, decay due to a negative transversal Lyapunov exponent, thus leading to stable synchronization.

Semiconductor lasers are of particular interest in the study of chaos synchronization. The synchronization properties may lead to new secure communication schemes [6–14]. However, as we will see in Chap. 10, it is impossible to chaos synchronize two delay coupled systems without self-feedback for large delays because the synchronized solution is always transversely unstable. In coupled lasers this effect leads to spontaneous symmetry breaking, and only generalized synchronization of leader-laggard type occurs [15].

However, chaos synchronization of two delay coupled systems can be stable (see Chap. 10) if each system has self-feedback. For semiconductor lasers this has been realized with a passive relay in form of a semitransparent mirror or an active relay in form of a third laser in between the two lasers [16–19]. These structures are thus interesting for chaos based communication systems.

V. Flunkert, *Delay-Coupled Complex Systems*, Springer Theses,
DOI: 10.1007/978-3-642-20250-6_9, © Springer-Verlag Berlin Heidelberg 2011

For practical applications it is not only necessary that the synchronized solution is (linearly) stable, but it is also important how robust the synchronization is to noise. Here, nonlinear effects may play an important role. In particular, for chaos synchronized systems bubbling [20, 21] plays a key role in this context. This effect may lead to occasional desynchronization even for arbitrarily small noise amplitudes. Bubbling has been observed for example in optical [22, 23] and electrical [24] systems.

This part is organized as follows. In Chap. 10 we analyze the stability of the synchronized solution in a general network of delay coupled units in the limit of large delay. We show that in this limit the master stability function (MSF) has a simple structure. From this symmetry we then draw general conclusions about the synchronizability of delay coupled systems.

In the rest of this part we investigate the synchronization properties of delay coupled lasers. After introducing the laser equations in Chap. 11, we consider two lasers coupled all-optically in different schemes in Chap. 12. Depending on the coupling topology necessary conditions on the delay times and coupling phases arise in order for the coupled systems to have a synchronized solution.

In Chap. 13, we will then investigate one of these coupling schemes (two bidirectionally delay-coupled lasers with delayed self-feedback) under the influence of noise. Here, we observe bubbling, which can be understood through the properties of the system's cavity modes.

References

1. A.S. Pikovsky, M.G. Rosenblum, J. Kurths, *Synchronization, A Universal Concept in Nonlinear Sciences*. (Cambridge University Press, Cambridge, 2001)
2. S. Boccaletti, J. Kurths, G. Osipov, D.L. Valladares, C.S. Zhou, The synchronization of chaotic systems. Phys. Rep. **366**, 1 (2002)
3. E. Mosekilde, Y. Maistrenko, D. Postnov, *Chaotic Synchronization: Applications to Living Systems*. (World Scientific, Singapore, 2002)
4. A.G. Balanov, N.B. Janson, D.E. Postnov, O.V. Sosnovtseva, *Synchronization: From Simple to Complex*. (Springer, Berlin, 2009)
5. L.M. Pecora, T.L. Carroll, Synchronization in chaotic systems. Phys. Rev. Lett. **64**, 821 (1990)
6. I. Fischer, Y. Liu, P. Davis, Synchronization of chaotic semiconductor laser dynamics on subnanosecond time scales and its potential for chaos communication. Phys. Rev. A **62**, 011801 (2000)
7. T. Heil, J. Mulet, I. Fischer, C.R. Mirasso, M. Peil, P. Colet, W. Elsäßer, On/off phase shift keying for chaos-encrypted communication using external-cavity semiconductor lasers. IEEE J. Quantum Electron. **38**, 1162 (2002)
8. A. Argyris, D. Syvridis, L. Larger, V. Annovazzi-Lodi, P. Colet, I. Fischer, J. García-Ojalvo, C.R. Mirasso, L. Pesquera, K.A. Shore, Chaos-based communications at high bit rates using commercial fibre-optic links. Nature **438**, 343 (2005)
9. D.M. Kane, K.A. Shore (eds), *Unlocking Dynamical Diversity: Optical Feedback Effects on Semiconductor Lasers*. (Weinheim, Wiley VCH, 2005)

10. W. Kinzel, I. Kanter, Secure communication with chaos synchronization. In: E. Schöll, H.G. Schuster (eds) *Handbook of Chaos Control, second completely revised and enlarged edition*, (Wiley-VCH, Weinheim, 2008)
11. R. Vicente, C.R. Mirasso, I. Fischer, Simultaneous bidirectional message transmission in a chaos-bases communication scheme. Opt. Lett. **32**, 403 (2007)
12. I. Kanter, E. Kopelowitz, W. Kinzel, Public channel cryptography: chaos synchronization and hilbert's tenth problem. Phys. Rev. Lett. **101**, 84102 (2008)
13. I. Kanter, Y. Aviad, I. Reidler, E. Cohen, M. Rosenbluh, An optical ultrafast random bit generator. Nat. Photon. **4**(1), 58–61 (2010)
14. W. Kinzel, A. Englert, I. Kanter, On chaos synchronization and secure communication. Phil. Trans. R. Soc. A **368**, 379 (2010)
15. J. Mulet, C.R. Mirasso, T. Heil, I. Fischer, Synchronization scenario of two distant mutually coupled semiconductor lasers. J. Opt. B **6**, 97 (2004)
16. I. Fischer, R. Vicente, J.M. Buldú, M. Peil, C.R. Mirasso, M.C. Torrent, J. García-Ojalvo, Zero-lag long-range synchronization via dynamical relaying. Phys. Rev. Lett. **97**, 123902 (2006)
17. E. Klein, N. Gross, M. Rosenbluh, W. Kinzel, L. Khaykovich, I. Kanter, Stable isochronal synchronization of mutually coupled chaotic lasers. Phys. Rev. E **73**, 066214 (2006)
18. L.B. Shaw, I.B. Schwartz, E.A. Rogers, R. Roy, Synchronization and time shifts of dynamical patterns for mutually delay-coupled fiber ring lasers. Chaos **16**, 015111 (2006)
19. A.S. Landsman, I.B. Schwartz, Complete chaotic synchronization in mutually coupled time-delay systems. Phys. Rev. E **75**, 026201 (2007)
20. P. Ashwin, J. Buescu, I. Stewart, Bubbling of attractors and synchronisation of chaotic oscillators. Phys. Lett. A **193**, 126 (1994)
21. E. Ott, J.C. Sommerer, Blowout bifurcations: the occurrence of riddled basins and on-off intermittency. Phys. Lett. A **188**, 39 (1994)
22. M. Sauer, F. Kaiser, On-off intermittency and bubbling in the synchronization break-down of coupled lasers. Phys. Lett. A **243**, 38 (1998)
23. J.R. Terry, K.S. Thornburg, D.J. DeShazer, G.D. VanWiggeren, S. Zhu, P. Ashwin, R. Roy, Synchronization of chaos in an array of three lasers. Phys. Rev. E **59**, 4036 (1999)
24. D.J. Gauthier, J.C. Bienfang, Intermittent loss of synchronization in coupled chaotic oscillators: Toward a new criterion for high-quality synchronization. Phys. Rev. Lett. **77**, 1751 (1996)

Chapter 10
Structure of the Master Stability Function for Large Delay

To determine the stability of a synchronized state in a network of identical units, a powerful method has been developed [1, 2] called the master stability function (MSF). Recent works [3–5] have started to investigate the MSF for networks with coupling delays. Time delay effects play an important role in realistic networks. For example, the finite propagation time of light between coupled semiconductor lasers [6–12] significantly influence the dynamics. Similar effects occur in neuronal [13–18] and biological [19] networks.

In this section we show [20] that the MSF has a very simple structure in the limit of large coupling delays. This allows us to prove a number of general statements about the synchronizability of networks with large coupling delay.

We will first discuss MSF-theory and since we are interested in delay coupled systems, we will do this in the context of networks with delay [3–5].

10.1 Approach

Consider a system of N identical units connected in a network with a coupling delay τ [4]

$$\frac{d}{dt}x^i(t) = f\left[x^i(t)\right] + \sum_{j=1}^{N} g_{ij}h\left[x^j(t - \tau)\right] \tag{10.1}$$

with $x^i \in \mathbb{R}^n$. Here, $g_{ij} \in \mathbb{R}$ is the coupling matrix determining the coupling topology and the strength of each link in the network, f is the (non-linear) function describing the dynamics of an isolated unit, and h is a possibly non-linear coupling function. A synchronized solution can only exist, if the row sum of the matrix is the same for each row, i.e., $\sigma = \sum_{j=1}^{N} g_{ij}$ independent of i. In this case if the systems start in a synchronized state, the feedback term will be equal for all x^i in

V. Flunkert, *Delay-Coupled Complex Systems*, Springer Theses,
DOI: 10.1007/978-3-642-20250-6_10, © Springer-Verlag Berlin Heidelberg 2011

(10.1) and there exists a synchronized solution. The synchronized solution $\bar{x}(t)$ is then determined by

$$\frac{d}{dt}\bar{x}(t) = f[\bar{x}(t)] + \sigma h [\bar{x}(t - \tau)]. \tag{10.2}$$

To calculate the stability of this synchronized solution, we consider small perturbations $\xi^i(t)$ to the individual systems

$$x^i(t) = \bar{x}(t) + \xi^i(t).$$

Inserting this ansatz into (10.1) and linearizing in ξ^i we find

$$\frac{d}{dt}\xi^i(t) = Df[\bar{x}(t)]\xi^i(t) + \sum_{j=1}^{N} g_{ij}Dh[\bar{x}(t - \tau)]\xi^i(t - \tau), \tag{10.3}$$

where Df and Dh are Jacobians. Using the vector

$$\Xi(t) = (\xi^1(t), \xi^2(t), \ldots, \xi^N(t))$$

this linear equation can be written as

$$\frac{d}{dt}\Xi(t) = I_N \otimes Df[\bar{x}(t)]\Xi(t) + g \otimes Dh[\bar{x}(t - \tau)]\Xi(t - \tau), \tag{10.4}$$

where I_N denotes the N-dimensional identity matrix. We can diagonalize the coupling matrix g with a unitary transformation U

$$\mathrm{diag}(\sigma, \gamma_1, \gamma_2, \ldots, \gamma_{N-1}) = UgU^{-1}.$$

Here, σ is the row sum of g, which is always an eigenvalue of g to the eigenvector $(1, 1, \ldots, 1)$. We call this the longitudinal eigenvalue of g. The other eigenvalues γ_k we then call the *transversal* eigenvalues of g.

Diagonalizing g in (10.4) with the transformation U does not affect the left hand side or the first term on the right hand side in (10.4), because both terms are multiples of the identity matrix in the subspace on which g acts. Thus after the diagonalization (10.4) is transformed into N equations

$$\frac{d}{dt}\xi(t) = Df[\bar{x}(t)]\xi(t) + \sigma Dh[\bar{x}(t - \tau)]\xi(t - \tau), \tag{10.5}$$

$$\frac{d}{dt}\xi(t) = Df[\bar{x}(t)]\xi(t) + \gamma_k Dh[\bar{x}(t - \tau)]\xi(t - \tau) \tag{10.6}$$

with $k = 1, \ldots, N - 1$. The first equation corresponds to perturbations in the direction of the vector $(1, 1, \ldots, 1)$, which act equally on each individual system and thus do not cause desynchronization. A growing perturbation in this direction indicates that the synchronized solution of the network is chaotic.

The $N-1$ other (10.6) on the other hand describe perturbations transversal to the SM. The synchronized solution is stable if and only if these perturbations decay, i.e., if the maximum Lyapunov exponent arising from the variational (10.6) is negative for all the transversal eigenvalues γ_k.

It was the idea of Pecora and Carroll [21] to define a function $\lambda_{\max}$, which maps a complex number $re^{i\phi}$ to the maximum Lyapunov exponent arising from the variational equation[1]

$$\frac{d}{dt}\xi(t) = Df[\bar{x}(t)]\xi(t) + re^{i\phi}Dh[\bar{x}(t-\tau)]\xi(t-\tau).$$

This function is called the master stability function and it can be calculated numerically. Once this is done on a sufficient domain in $\mathbb{C}$, we can immediately decide for any network structure, whether synchronization will be stable or not. We only need to evaluate the MSF at the transversal eigenvalues γ_k of the particular network's coupling matrix. This way the problem has been separated into a part, which only depends on the dynamics of the individual system, and a part which only depends on the coupling topology.

We will now restrict our analysis to maps [4], but all ingredients of our argument are also valid for flows and we will point out, where the results differ slightly for flows. Delay coupled maps have been widely studied because they show similar behavior as DDEs and interesting synchronization phenomena have been found in these systems [22].

For delay coupled maps the dynamics in the SM is governed by the equation $x_{k+1} = f(x_k) + \sigma h(x_{k-\tau})$ with $\tau \in \mathbb{N}$ and $x_k \in \mathbb{C}^d$ or $\in \mathbb{R}^d$ and the MSF is calculated from

$$\xi_{k+1} = Df(x_k)\xi_k + re^{i\psi}Dh(x_{k-\tau})\xi_{k-\tau}. \tag{10.7}$$

Whether the synchronized dynamics is chaotic or not depends on whether the MSF evaluated at the eigenvalue $re^{i\psi} = \sigma$, which corresponds to perturbations parallel to the SM, is positive or not.

With the matrix coefficients $A_k := Df(x_k)$, and $B_k := Dh(x_{k-\tau})$ the variational equation is given by

$$\xi_{k+1} = A_k\xi_k + re^{i\psi}B_k\xi_{k-\tau}. \tag{10.8}$$

Note that when the delay is changed the dynamics in the SM changes, too. Hence, we are not able to make predictions about what happens as τ is changed. However, at a fixed large value of the delay time τ we can compare the Lyapunov exponents arising from different values of $re^{i\psi}$ in (10.7).

We will now analyze the Lyapunov exponents arising from (10.8) in the limit of large τ. We do this in the following steps: first we analyse the two simpler cases, where the dynamics in the SM is a fixed point FP or a PO. Then to expand the

---

[1] Note that the complex number $re^{i\phi}$ is usually denoted by $\alpha + i\beta$ in the literature.

results to chaotic trajectories x_k in the SM we use the fact that PO are dense in a chaotic attractor.

10.2 A Fixed Point in the Synchronization Manifold

For FPs of delay differential equations there exists a scaling theory for the FP's eigenvalues in the limit of large delay [23–26]. Recently this theory has been generalized to the scaling of Floquet exponents [27]. In both cases the eigenvalues or Floquet spectrum consist of two parts: a strongly unstable part arising from unstable eigenvalues of the system without delay and a pseudo-continuous spectrum, for which the real part of the eigenvalues approach zero in the limit of large delay. This scaling theory has been developed for flows. Since we restrict ourselves to maps, we want to discuss the scaling theory for maps now. However, each step can be done in the same way for flows by applying the large delay theory developed in [23–27].

Let us first consider the case, where the dynamics in the SM is a FP, i.e., a period $T = 1$ orbit. In this case the coefficient matrices in (10.8) are constant $A = A_k$ and $B = B_k$.

Making the ansatz $\xi_k = z^k \xi_0$, we find an equation for the multipliers z

$$\chi(z) = \det[A - zI + re^{i\psi}Bz^{-\tau}] = 0, \tag{10.9}$$

where I denotes the identity-matrix.

For the strongly unstable spectrum we suppose there is a solution with $|z| > 1$. Then in the limit of $\tau \to \infty$ (10.9) becomes

$$\det[A - zI] = 0. \tag{10.10}$$

Thus in the limit of large delay the eigenvalues z of A with $|z| > 1$ are also solutions of (10.9).

We are now interested in the pseudo-continuous spectrum, i.e., in the solutions with $|z| \approx 1$ in the limit of large τ. We make the ansatz $z = (1 + \delta/\tau)e^{i\omega}$. In the limit $\tau \to \infty$ we have $(1 + \delta/\tau)^{-\tau} \to e^{-\delta}$, and $(1 + \delta/\tau) \to 1$. Thus in the limit $\tau \to \infty$ (10.9) becomes

$$0 = \det[A - Ie^{i\omega} + re^{-\delta}e^{i(\psi-\phi)}B] \tag{10.11}$$

with $\phi = \omega\tau$. Since the curve parameter ϕ on the branch takes on any (arbitrarily dense) value in $[0, 2\pi]$, we can already see that the phase ψ in the variational equation does not change δ, i.e., the MSF is invariant under phase shifts (rotations) and its value only depends on r.

If B is invertible, we can calculate the eigenvalues $\mu = re^{-\delta}e^{i(\psi-\phi)}$ in the following equation

$$0 = \det[-B^{-1}(A - Ie^{i\omega}) - \mu], \tag{10.12}$$

which is a polynomial in μ. This polynomial has exactly d roots μ_j $(j = 1, \ldots, d)$, which are eigenvalues of $-B^{-1}(A - Ie^{i\omega})$.

If B is not invertible, (10.11) still gives a polynomial in μ, for which the roots can be calculated. Then each eigenvalue μ will be a function of ω and one can find the branches

$$\delta(\omega) = \ln\left[\frac{r}{|\mu(\omega)|}\right] = -\ln|\mu(\omega)| + \ln r.$$

The function $\mu(\omega)$ can admit the zero value at some point ω_0, i.e., $\mu(\omega_0) = 0$, in the case when the matrix A has an eigenvalue with $|z| = 1$. Indeed, as follows from (10.12), for $\mu = 0$, $\omega = \omega_0$ and $\det B \neq 0$ we have

$$\det[A - Ie^{i\omega_0}] = \det[A - Iz] = 0.$$

In all other cases, with $\det B \neq 0$ and $|z| \neq 1$, the function $|\mu(\omega)|$ is bounded $0 < \mu_0 \leq |\mu(\omega)| \leq \mu_1$.

If there are no strongly unstable eigenvalues, the sign of δ determines the stability in the limit of large τ. It is clear, that δ increases monotonically with increasing r and in particular δ is negative for small r and positive for large r. Thus there is a minimum radius r_0, for which the first eigenvalue branch becomes unstable $\delta > 0$ and thus the MSF changes sign.

Note that we have obtained the function $\delta(\omega)$ on which the solutions lie in the limit of large τ but not yet the exact values of ω. These are not important in the limit of large delay, since the eigenvalues become very dense on the curve $\delta(\omega)$. Indeed, the exact values of ω can be calculated from the expression $\mu(\omega) = re^{-\delta(\omega)}e^{i(\psi - \omega\tau)}$, which implies

$$\text{Arg}\,\mu(\omega) = \psi - \omega\tau + 2\pi k \qquad (10.13)$$

for any integer k. Since $\mu(\omega)$ is a known eigenvalue of the matrix $-B^{-1}(A - Ie^{i\omega})$, (10.13) can be considered as a transcendental equation for determining the solutions $\omega = \omega_k$. In particular, (10.13) implies that the distance between the neighboring solutions ω_k and ω_{k-1} reads

$$\omega_k - \omega_{k-1} = \frac{1}{\tau}[\text{Arg}\,\mu(\omega_{k-1}) - \text{Arg}\,\mu(\omega_k)] + \frac{2\pi}{\tau}$$
$$= 2\pi/\tau + \mathcal{O}(1/\tau^2).$$

Thus it is proportional to $1/\tau$ and the curve $\delta(\omega)$ is filled densely with roots as $\tau \to \infty$. Note that the curve $\delta(\omega)$ is determined in the bounded interval $\omega \in [0, 2\pi]$ in contrast to the case of DDEs [24], where ω was varying on the whole axis $(-\infty, \infty)$.

The simple case is that of a one dimensional $(d = 1)$ complex map, where $A = a \in \mathbb{C}$ and $B = b \in \mathbb{C}$ with $|a| < 1$. In this case we can explicitly calculate

Fig. 10.1 Pseudo continuous spectrum $\delta(\omega)$ (*lines*) and location of the exact roots (*crosses*) for the example of a one dimensional complex map for $r = 3.3 > r_0 = 3$ and $r = 2.7 < r_0 = 3$. Parameters: $a = 0.4, b = 0.2, \psi = 0,$ $\tau = 30$

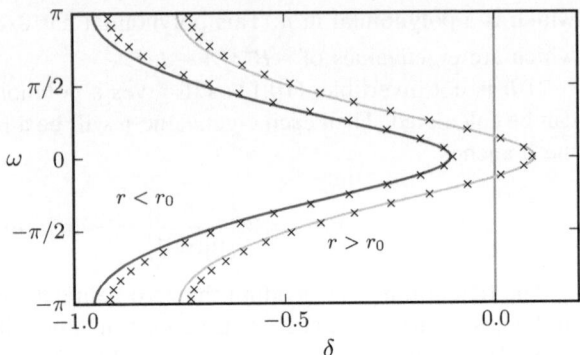

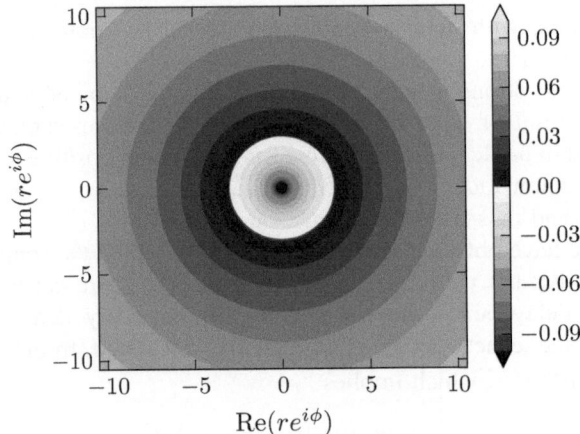

Fig. 10.2 Master stability function for a one dimensional map with FP-dynamics in the SM. The *red* regions correspond to $\lambda_{max} > 0$ (synchronized state is unstable). The *gray* regions correspond to $\lambda_{max} < 0$ (synchronized state is stable). The *blue* circle indicates the stability boundary given by r_0 according to (10.14). Already for relatively low values of τ the *blue line* matches very well the numerically obtained boundary. Parameters of the variational equation: $a = 0.4, b = 0.2, \tau = 20$

$$\delta(\omega) = \ln(|rb|/|a - e^{i\omega}|).$$

For $r < (1 - |a|)/|b|$ all the eigenvalues approach magnitude 1 from the stable side and for $r > (1 - |a|)/|b|$ there are always weakly unstable eigenvalues. Thus the MSF changes sign at

$$r_0 = (1 - |a|)/|b|. \tag{10.14}$$

The pseudo-continuous spectrum for these two cases is depicted in Fig. 10.1. The corresponding MSF is shown in Fig. 10.2. As $\tau \to \infty$, the $\lambda_{max} = 0$ contour line approaches the circle with radius r_0. This is depicted in Fig. 10.3, where the

Fig. 10.3 Boundary $r(\psi)$ of
stability domain in polar
coordinates for different
values of τ in logarithmic
scale. With increasing delay
$r(\psi) \to r_0$. Parameters of the
variational equation:
$a = 0.4, b = 0.2$

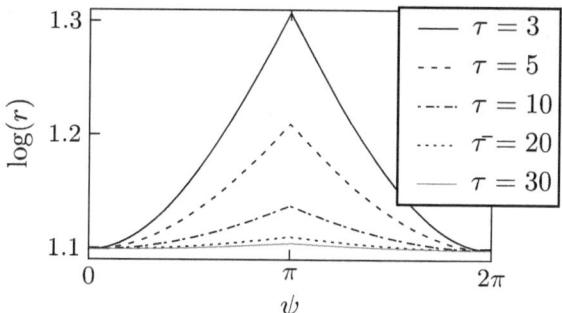

angle-dependency of the critical radius is shown for different values of τ in a
logarithmic scale. For small values of τ, the critical radius has a strong angle-
dependency. However, already for a value of $\tau = 20$, the rotation symmetry is
almost perfect (see Fig. 10.2).

10.3 A Periodic Orbit in the Synchronization Manifold

Now consider the map

$$\xi_{k+1} = A_k\xi_k + re^{i\psi}B_k\xi_{k-\tau}, \qquad (10.15)$$

where A_k and B_k are periodic with period T, corresponding to a PO in the SM. We
consider the case of large delay, i.e., $\tau > T$.

Making a Floquet–like ansatz $\xi_k = z^k q_k$, where q_k is T periodic we find

$$z\,q_{k+1} = A_k\,q_k + re^{i\psi}B_k\,z^{-\tau}q_{k-n} \qquad (10.16)$$

with $n = \tau \bmod T \in \{0, 1, \ldots, T-1\}$.

For the strongly unstable spectrum again suppose there is a solution with
$|z| > 1$, then in the limit $\tau \to \infty$ the term $z^{-\tau}$ vanishes and we find

$$z\,q_{k+1} = A_k\,q_k. \qquad (10.17)$$

Using the periodicity of q_k, (10.17) implies

$$\det[z^T - \prod_{k=1}^{T} A_k] = 0,$$

where z^T is a Floquet multiplier of the system $\xi_{k+1} = A_k\xi_k$ without delay. Hence,
if z^T is a Floquet multiplier of (10.17), with $|z| > 1$, then in the limit $\tau \to \infty$ it is
also a solution of (10.15) and vice versa.

For the pseudo-continuous spectrum we again make the ansatz $z = (1 + \delta/\tau)e^{i\omega}$ and taking the limit $\tau \to \infty$ (10.16) becomes

$$e^{i\omega}q_{k+1} = A_k\,q_k + re^{-\delta}e^{i(\psi-\phi)}B_k\,q_{k-n} \qquad (10.18)$$

with $\phi = \omega\tau$. One thus has to solve

$$0 = [e^{i\omega}\overline{J} + \overline{A} + \mu\overline{B}]\mathbf{q} = 0, \qquad (10.19)$$

where $\overline{A} = \mathrm{diag}\{A_1, \ldots, A_T\}$,

$$\overline{J} = \begin{bmatrix} 0 & & I \\ I & \ddots & \\ & \ddots & \ddots \\ & & I & 0 \end{bmatrix}, \quad \overline{B} = \begin{bmatrix} 0 & & B_1 & & \\ & \ddots & & \ddots & \\ & & \ddots & & B_n \\ B_{n+1} & & \ddots & & \\ & \ddots & & & \\ & & B_T & & 0 \end{bmatrix},$$

$\mu = re^{-\delta}e^{i(\psi-\phi)}$, and $\mathbf{q} = (q_1, \ldots, q_T)$. The position of the diagonal lines in the matrix $\overline{B}$ depends on the value of $n = \tau \bmod T$. Taking the determinant of the matrix in (10.19) results in a polynomial in $\mu = re^{-\delta}e^{i(\psi-\phi)}$ (of maximum order $d \times T$). Again, the roots μ are functions of ω and we can calculate the branches $\delta(\omega) = -\ln|\mu(\omega)| + \ln r$, where ψ and ϕ drop out. As in the case of FPs, one can show that the function $|\mu(\omega)|$ is bounded $0 < \mu_0 \le |\mu(\omega)| \le \mu_1$ unless the instantaneous system has a Floquet multiplier z with $|z| = 1$.

We have again found the same structure of the MSF : The MSF is rotationally symmetric in the complex plane about the origin. If without feedback ($r = 0$) the MSF is positive, then it is constant in the limit of large delay. Otherwise it is a monotonically increasing function of r and there is a critical radius r_0 where it changes sign.

10.4 A Chaotic Attractor in the Synchronization Manifold

In every chaotic attractor there is an infinite number of UPOs embedded. It has been long known that the characteristic properties of the chaotic system can be described in terms of these PO [28]. Intuitively, the chaotic trajectory follows the UPOs closely and "switches" between them, thus averaging over the UPOs in the appropriate way allows us to calculate statistical properties of the attractor. One of the most important examples is the natural measure of the chaotic attractor, which is concentrated at the UPO (hot-spots) and can in fact be expressed in terms of the orbit's Floquet multipliers [29, 30].

Lyapunov exponents arising from variational equations such as (10.8) have been discussed in the framework of periodic orbit theory [28, 31–33], too. In particular it has been shown [34] that a chaotic attractor in an invariant manifold loses its transversal stability in a blow-out bifurcation when the transversely unstable orbits outweigh the transversely stable orbits. To be precise, we divide the orbits into two groups of transversely stable and unstable orbits and define [34] the transversely stable weight Λ_T^s and the unstable weight Λ_T^u as

$$\Lambda_T^{u,s} = \sum_{j=1}^{N_T^{u,s}} \mu_T(j) \lambda_T(j), \qquad (10.20)$$

where the sum goes over all N_T^u transversely unstable and N_T^s transversely stable orbits of period T, respectively. Here, $\mu_T(j)$ is the weight of the j-th orbit, corresponding to the natural measure of a typical trajectory in the neighborhood of the j-th orbit and $\lambda_T(j)$ is the transversal Lyapunov exponent of this j-th orbit. The weight of a PO is inversely proportional to the product of its unstable Floquet multipliers [29, 33]. The attractor is transversely unstable if and only if in the limit of large T

$$\Lambda_T^u > |\Lambda_T^s|. \qquad (10.21)$$

We now make the connection to the scaling theory for large τ. Starting from $r = 0$ (no feedback) transversal Lyapunov exponents $\lambda_T(j)$ of each orbit can only increase with increasing r, as shown above. In particular for large enough r the orbits become transversely unstable: either they are already unstable for $r = 0$ and thus remain unstable or the pseudo-continuous spectrum goes to zero and for large r it does so from the unstable side. Thus there exists a minimum radius r_0, for which the condition (10.21) on the weights is fulfilled. Note that since we consider the limit $\tau \to \infty$ we can evaluate (10.21) at arbitrarily large T, although it is a common result of PO theory that formulas such as (10.20) converge quickly.

Thus in summary the MSF has the same structure as for FPs and POs (the rotation symmetry follows from the rotation symmetry of each $\lambda_T(j)$).

10.5   Consequences for Synchronization of Networks

Let us now discuss what the structure of the MSF means for the synchronizability of networks. We can categorize networks into three types depending on the magnitude of the largest transversal eigenvalue $\gamma_{\max}$ in relation to the magnitude of row sum σ: (A) the largest transversal eigenvalue is strictly smaller than the magnitude of the row sum ($|\gamma_{\max}| < |\sigma|$), (B) the largest transversal eigenvalue has the same magnitude as the row sum ($|\gamma_{\max}| = |\sigma|$), and (C) the largest transversal eigenvalue has a larger magnitude than the row sum ($|\gamma_{\max}| > |\sigma|$).

At $r = |\sigma|$ the MSF is positive ($r_0 < |\sigma|$) for chaotic dynamics in the SM and negative ($|\sigma| < r_0$) for dynamics on a stable PO or a FP. This gives us a lower or an upper bound on r_0 and we can thus give the classification as shown in Table 10.1.

Table 10.1 Stability of chaotic and non-chaotic synchronized solutions for the three types of networks

		Chaotic dynamics in the SM $(r_0 < |\sigma|)$	PO or FP in the SM $(|\sigma| < r_0)$
(A)	$|\gamma_{max}| < |\sigma|$	Synchr. stable if $|\gamma_{max}| < r_0$	Synchr. stable
(B)	$|\gamma_{max}| = |\sigma|$	Synchr. unstable	Synchr. stable
(C)	$|\gamma_{max}| > |\sigma|$	Synchr. unstable	Synchr. stable if $|\gamma_{max}| < r_0$

In networks of type (A) and (B) synchronization on a FP or a PO, which is stable within the SM, is always stable. For type (C) this dynamics may be stable or not depending on the particular network (value of $|\gamma_{max}|$) and the dynamics in the SM (value of r_0). On the other hand chaos synchronization is always unstable in networks of type (B) and (C) and it may be stable or not in networks of type (A) again depending on the particular network and the dynamics.

Note that for autonomous flows with a stable PO in the SM we always have $r_0 = |\sigma|$, due to the PO's Goldstone mode. Thus for this case we cannot decide whether synchronization for type (B) networks will be stable or not. This depends on whether the $\lambda_{max} = 0$ contour line of the MSF approaches the circle with radius $r_0 = |\sigma|$, locally, at the transversal eigenvalues with $|\gamma_k| = |\sigma|$, from the outside (stable) or from the inside (unstable).

We now list some examples for the three types of networks. The categorization follows from the eigenvalue structure (spectral radius) for the corresponding matrices, which can, for instance, be derived using Gerschgörin's theorem.

- Mean field coupled systems have $\gamma_k = 0$ for all K and are thus of type (A).
- Networks with only inhibitory connections (negative entries) or only excitatory connections (positive entries) are up to the row sum factor stochastic matrices, i.e., the coupling matrix G can be written as

$$G = \sigma P,$$

where P is a stochastic matrix (positive entries and row sum one). For stochastic matrices it is well known that the spectral radius is one, i.e., all eigenvalues have magnitude smaller than or equal to one. The proof utilizes Gerschgörin's theorem [35]. Thus it follows for G that no eigenvalues has magnitude larger than $|\sigma|$ and these networks are of type (A) or (B).

- Rings of uni-directionally coupled elements and two bidirectionally coupled elements are of type (B) [5].
- Any network with zero row sum ($\sigma = 0$) is of type (B) (trivial case) or (C).
- Two bidirectionally coupled systems without self-feedback are of type (B).

In the literature there is a great amount of material on the relation of the spectral radius and the row sum for certain types of matrices. These results are immediately applicable to our classification. For a concrete network topology the classification is of course very simple.

Fig. 10.4 Master stability function (shown in *color*) for delay coupled logistic maps (10.23). Panel (a) corresponds to $\lambda = 3.2$ and the stable period-2 orbit within the SM. Panel (b) corresponds to $\lambda = 3.8$ and a chaotic attractor within the SM. In both cases the delay is chosen as $\tau = 30$

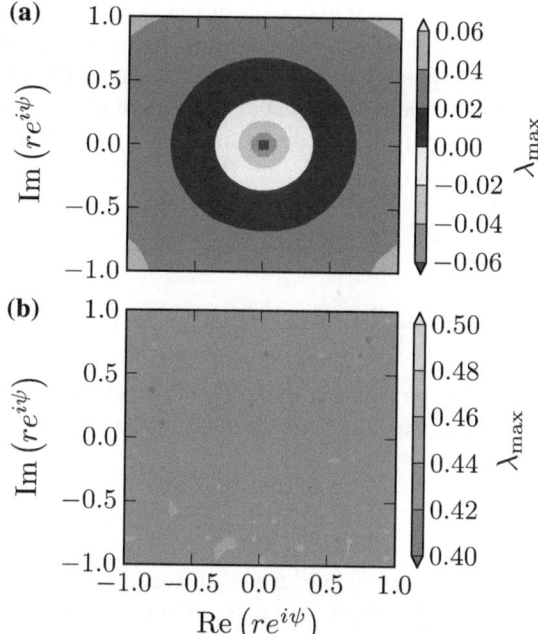

Networks with $\sigma = 0$ belong to class (B) (trivial case) or to class (C). This confirms the conjecture stated in [4]: Networks for which the trajectory of an uncoupled unit is also a solution of the network ($\sigma = 0$) cannot exhibit chaos synchronization for large coupling delay.

For the chaotic case there may exist another radius r_b, with $0 \leq r_b \leq r_0$, where the first PO in the attractor loses its transverse stability and the attractor undergoes a bubbling bifurcation [12, 36, 37] (see Chap 13). Then any network with $r_b < |\gamma_{max}| < r_0$ will exhibit bubbling in the presence of noise (or parameter mismatch), while any network with $|\gamma_{max}| < r_b$ will show stable synchronization, even in the presence of noise.

In order to illustrate the obtained results, let us consider the following example of linearly coupled logistic maps

$$x^m_{k+1} = \lambda x^m_k (1 - x^m_k) + \sum_{j=1}^{N} g_{mj} x^j_{k-\tau} \tag{10.22}$$

with the zero row sums $\sigma = 0$. The MSF is calculated from the following delayed system

$$\xi_{k+1} = \lambda(1 - 2x_k)\xi_k + re^{i\psi}\xi_{k-\tau}, \tag{10.23}$$

where the dynamics on the synchronization manifold x_k is determined by the map $x_{k+1} = \lambda x_k (1 - x_k)$. Figure. 10.4 shows numerically computed MSF, i.e., the largest Lyapunov exponent of the system (10.23) for two different cases: $\lambda = 3.2$ and $\lambda = 3.8$, which correspond to the stable period-2 state and chaos, respectively. The delay is set to $\tau = 30$. In both cases, the MSF are radially symmetric. In the stable periodic case (panel (a)), there exists a critical radius r_0 where the MSF changes sign, which determines the synchronizability of a given coupled system. In the chaotic case (panel (b)) the MSF is close to a positive constant, i.e., any coupling configuration will be unstable.

10.6 Experimental Setup for Finding the Critical Radius

We now propose an experimental method for determining the critical radius r_0. Consider two elements coupled in the following network motif

$$x_{k+1}^1 = f(x_k^1) + \mu h(x_{k-\tau}^1) + \nu h(x_{k-\tau}^2),$$
$$x_{k+1}^2 = f(x_k^2) + \mu h(x_{k-\tau}^2) + \nu h(x_{k-\tau}^1),$$

where μ and ν are the self feedback strengths and the coupling strengths, respectively. Suppose we are able to change the self-feedback strengths μ and the coupling strengths ν, for example by using gray filters in an optical experiment.

Let us choose

$$\mu = \frac{1}{2}(\sigma + r) \quad \text{and} \quad \nu = \frac{1}{2}(\sigma - r).$$

Then the dynamics in the SM is given by

$$x_{k+1} = f(x_k) + \sigma h(x_{k-\tau}),$$

while the variational equation transverse to the SM is given by

$$\xi_{k+1} = Df(x_k)\xi_k + r Dh(x_{k-\tau})\xi_{k-\tau}.$$

Thus by changing r (for fixed σ) and checking whether the two elements synchronize we are able to probe the MSF along the real axis at the radius r. Due to the monotonicity we can use a root-finding algorithm such as the bisection method to find r_0 to high accuracy with little iterations of the experiment and without knowledge of the functions f and h. We can repeat this procedure for other values of σ and obtain the critical radius as a function of $r_0(\sigma)$. Thus from this rather simple setup we can decide for any network of these elements whether synchronization is stable or not.

As an example we consider two optoelectronically coupled lasers (see Sect. 7.2)

Fig. 10.5 Schematic setup for determining the critical radius in an experiment

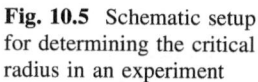

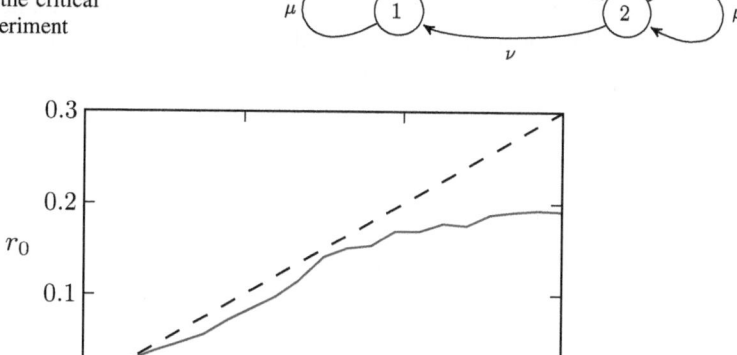

Fig. 10.6 Numerically calculated critical radius r_0 as a function of σ (*solid curve*) for the system of optoelectronically coupled lasers corresponding to 10.24. A network can only have a stable synchronized solution if the magnitudes of its transversal eigenvalues are below the curve. The curve is calculated up to an absolute error of 10^{-4}. The dashed line shows the diagonal line $r_0 = \sigma$. Parameters: $\varepsilon_0 = 10^{-7}$, $p = 1, T = 200, \tau = 2000$

$$\frac{d}{dt}\rho_1 = n_1\rho_1,$$
$$T\frac{d}{dt}n_1 = p + \mu\rho_1(t-\tau) + v\rho_2(t-\tau) - n_1 - (1+n_1)\rho_1, \tag{10.24a}$$

$$\frac{d}{dt}\rho_2 = n_2\rho_2,$$
$$T\frac{d}{dt}n_2 = p + \mu\rho_2(t-\tau) + v\rho_1(t-\tau) - n_2 - (1+n_2)\rho_2, \tag{10.24b}$$

where ρ_i and n_i is the intensity and the carrier density of the ith laser, respectively. The pump current of each laser is modulated by the delayed intensities according to the coupling scheme depicted in Fig. 10.5. As discussed in Sect. 7.2 such feedback can be realized by using photodiodes to measure the intensities of the arriving signals and modulating the pump current accordingly. The bidirectional coupling has strength v and the self-feedback of each laser has strength μ. For this setup we have numerically calculated $r_0(\sigma)$ in the same manner as it would be done in an experiment: We choose a value of σ, an interval $I_r = [r_{min}, r_{max}]$ for the r-domain and an initial value of $r = r_0$. We consider the systems to be synchronized, if the relative synchronization error

$$\varepsilon := \frac{\langle|\rho_1 - \rho_2|\rangle}{\frac{1}{2}[\langle\rho_1\rangle + \langle\rho_2\rangle]}$$

is smaller than a threshold ε_0. We then simulate the system and use the bisection method to find the synchronization threshold r_0 (up to a desired accuracy) in the interval I_r. We can then use the calculated value of r_0 as an initial guess for neighboring σ-values and thus follow the curve $r_0(\sigma)$.

The result is depicted in Fig. 10.6, where the solid curve shows $r_0(\sigma)$ and the dashed line corresponds to $r_0 = \sigma$. For small values of σ, i.e., weak feedback, the dynamics is a PO and due to the Goldstone mode we have $r_0 \approx \sigma$. For larger values of σ, the system becomes chaotic and $r_0 < \sigma$. For a given value of σ, a network has a stable synchronized solution if and only if all transversal eigenvalues γ_k of the corresponding coupling matrix have magnitude $|\gamma_k| < r_0(\sigma)$.

10.7 Conclusion and Outlook

In conclusion we have shown that the MSF has a simple structure in the limit of large delay: it is rotationally symmetric around the origin and either positive and constant (if it is positive at the origin) or monotonically increasing and becomes positive at a minimum radius r_0. This structure allowed us to prove a recent conjecture [4] about synchronizability of chaotic elements. Furthermore, we classified network structures into three types depending on the magnitude of the maximum transversal eigenvalue in relation to the magnitude of the row sum and showed that these network types have distinct synchronization properties.

The rotational symmetry of the MSF have previously been found numerically [5, 35]. In Ref. [35] the same structure of the MSF has been found for a PO in the SM for which the period T is approximately equal to the delay time τ. For this case the structure of the MSF has also been derived analytically in [35]. Note that this case is complementary to the situation $T \ll \tau$ that we looked at in this section. So the structure of the MSF that we found seems to be valid in even more general cases.

The derived results are only valid for networks, where each link has the same coupling delay. Recently, however, an interesting observation has been made when two systems are coupled via multiple delays [38]. As we saw above two bidirectionally delay coupled systems without self-feedback can not synchronize in the limit of large delay (the network is of type (B)).

In Ref. [38] A. Englert et al. showed that zero-lag synchronization can be stable without self-feedback, if the two systems are bidirectionally coupled via multiple delays. It may thus be interesting to consider networks with two (or more) delays of the form

$$x_{k+1}^i = f(x_k^i) + \sum_{j=1}^{N} g_{ij}^{(1)} h^{(1)}(x_{k-\tau_1}^j) + \sum_{j=1}^{N} g_{ij}^{(2)} h^{(2)}(x_{k-\tau_2}^j),$$

where $g^{(1)}$ and $g^{(2)}$ are the two coupling matrices and $h^{(1)}$ and $h^{(2)}$ are the coupling functions corresponding to the two types of links with two different delays.

If the coupling matrices commute $[g^{(1)}, g^{(2)}] = 0$, they can be diagonalized simultaneously and one would obtain a MSF of two arguments

$$\lambda_{\max}(r_1 e^{i\psi_1}, r_2 e^{i\psi_2})$$

corresponding to the variational equation

$$\xi_{k+1} = Df(x_k)\xi_k + r_1 e^{i\psi_1} Dh^{(1)}(x_{k-\tau_1})\xi_{k-\tau_1} + r_2 e^{i\psi_2} Dh^{(2)}(x_{k-\tau_2})\xi_{k-\tau_2}.$$

Then it would be interesting to consider the limit $\tau_1 \to \infty$, $\tau_2 \to \infty$ with a fixed ratio $R = \tau_1/\tau_2$. The asymptotic analysis for large delay times has not been applied to multiple delays yet, and it would be interesting to investigate this theory in this context. Depending on whether the ratio is rational (of low order, e.g.,$1 : 1, 1 : 2, 2 : 3, \ldots$) or irrational we can expect different synchronization properties [16, 38].

On the other hand, if $[g^{(1)}, g^{(2)}] \neq 0$, the MSF approach fails, because the two matrices cannot be diagonalize simultaneously. Note that this is already the case without delay but with two different coupling functions $h^{(1)}$ and $h^{(2)}$, i.e., a network with two types of links. It is not clear how a MSF approach could be generalized to this situation.

References

1. L.M. Pecora, T.L. Carroll, Master stability functions for synchronized coupled systems. Phys. Rev. Lett. **80**, 2109 (1998)
2. L.M. Pecora, M. Barahona, Synchronization of Oscillators in Complex Networks. in *New Research on Chaos and Complexity*, chap 5, ed. by F.F. Orsucci, N. Sala, (Nova Science Publishers, Hauppauge, 2006), pp. 65–96
3. M. Dhamala, V.K. Jirsa, M. Ding, Enhancement of neural synchrony by time delay. Phys. Rev. Lett. **92**, 074104 (2004)
4. W. Kinzel, A. Englert, G. Reents, M. Zigzag, I. Kanter, Synchronization of networks of chaotic units with time-delayed couplings. Phys. Rev. E. **79**, 056207 (2009)
5. C.-U. Choe, T. Dahms, P. Hövel, E. Schöll, Controlling synchrony by delay coupling in networks from in-phase to splay and cluster states. Phys. Rev. E. **81**, 025205(R) (2010)
6. H. Erzgräber, B. Krauskopf, D. Lenstra, Compound laser modes of mutually delay-coupled lasers. SIAM J. Appl. Dyn. Syst. **5**, 30 (2006)
7. W. Carr, I.B. Schwartz, M.Y. Kim, R. Roy, Delayed-mutual coupling dynamics of lasers: scaling laws and resonances. SIAM J. Appl. Dyn. Syst. **5**, 699 (2006)
8. O. D'Huys, R. Vicente, T. Erneux, J. Danckaert, I. Fischer, Synchronization properties of network motifs: Influence of coupling delay and symmetry. Chaos. **18**, 037116 (2008)
9. I. Fischer, R. Vicente, J.M. Buldú, M. Peil, C.R. Mirasso, M.C. Torrent, J. García-Ojalvo, Zero-lag long-range synchronization via dynamical relaying. Phys. Rev. Lett. **97**, 123902 (2006)
10. R. Vicente, L.L. Gollo, C.R. Mirasso, I. Fischer, P. Gordon, Dynamical relaying can yield zero time lag neuronal synchrony despite long conduction delays. Proc. Natl. Acad. Sci. **105**, 17157 (2008)

11. H.J. Wünsche, S. Bauer, J. Kreissl, O. Ushakov, N. Korneyev, F. Henneberger, E. Wille, H. Erzgräber, M. Peil, W. Elsäßer, I. Fischer, Synchronization of delay-coupled oscillators: A study of semiconductor lasers. Phys. Rev. Lett. **94**, 163901 (2005)
12. V. Flunkert, O. D'Huys, J. Danckaert, I. Fischer, E. Schöll, Bubbling in delay-coupled lasers. Phys. Rev. E. **79**, 065201 (2009)
13. E. Rossoni, Y. Chen, M. Ding, J. Feng, Stability of synchronous oscillations in a system of Hodgkin-Huxley neurons with delayed diffusive and pulsed coupling. Phys. Rev. E. **71**, 061904 (2005)
14. C. Hauptmann, O. Omel'chenko, O.V. Popovych, Y. Maistrenko, P.A. Tass, Control of spatially patterned synchrony with multisite delayed feedback. Phys. Rev. E. **76**, 066209 (2007)
15. C. Masoller, M.C. Torrent, J. García-Ojalvo, Interplay of subthreshold activity, time-delayed feedback, and noise on neuronal firing patterns. Phys. Rev. E. **78**, 041907 (2008)
16. E. Schöll, G. Hiller, P. Hövel, M.A. Dahlem, Time-delayed feedback in neurosystems. Phil. Trans. R. Soc. A. **367**, 1079 (2009)
17. M.A. Dahlem, M.H. Frank, W. Dobler, E. Schöll, (2008) Curvature-induced stabilization of particle-like waves. in prep. for Phys. Rev. Let.
18. P. Hövel, M.A. Dahlem,T. Dahms, G. Hiller,E. Schöll, (2009) Time-delayed feedback control of delay-coupled neurosystems and lasers, in Preprints of the Second IFAC meeting related to analysis and control of chaotic systems (CHAOS09) (World Scientific, Singapore), (arXiv:0912.3395)
19. A. Takamatsu, R. Tanaka, H. Yamada, T. Nakagaki, T. Fujii, I. Endo, Spatiotemporal symmetry in rings of coupled biological oscillators of physarum plasmodial slime mold. Phys. Rev. Lett. **87**, 078102 (2001)
20. V. Flunkert, S. Yanchuk, T. Dahms, E. Schöll, Synchronizing distant nodes: a universal classification of networks. Phys. Rev. Lett. **105**, 254101 (2010)
21. L.M. Pecora, T.L. Carroll, Synchronization in chaotic systems. Phys. Rev. Lett. **64**, 821 (1990)
22. F.M. Atay, J. Jost, A. Wende, Delays, connection topology, and synchronization of coupled chaotic maps. Phys. Rev. Lett. **92**, 144101 (2004)
23. G. Giacomelli, A. Politi, Relationship between delayed and spatially extended dynamical systems. Phys. Rev. Lett. **76**, 2686 (1996)
24. S. Yanchuk, M. Wolfrum (2005) Instabilities of equilibria of delay-differential equations with large delay, in Proceeding of the 5th EUROMECH Nonlinear Dynamics Conference ENOC-2005, Eindhoven, edited by D. H. van Campen, M. D. Lazurko, W. P. J. M. van den Oever (Eindhoven University of Technology, Eindhoven, Netherlands), pp. 1060–1065, eNOC Eindhoven (CD ROM), ISBN 90 386 2667 3
25. S. Yanchuk, M. Wolfrum, P. Hövel, E. Schöll, Control of unstable steady states by long delay feedback. Phys. Rev. E. **74**, 026201 (2006)
26. M. Wolfrum, S. Yanchuk, Eckhaus instability in systems with large delay. Phys. Rev. Lett. **96**, 220201 (2006)
27. S. Yanchuk, P. Perlikowski, Delay and periodicity. Phys. Rev. E. **79**, 046221 (2009)
28. P. Cvitanović, R. Artuso, R. Mainieri, G. Tanner, G. Vattay, *Chaos: Classical and Quantum*. (Niels Bohr Institute, Copenhagen, 2008) http://ChaosBook.org
29. C. Grebogi, E. Ott, J.A. Yorke, Unstable periodic orbits and the dimensions of multifractal chaotic attractors. Phys. Rev. A. **37**, 1711 (1988)
30. Y.C. Lai, Y. Nagai, C. Grebogi, Characterization of the natural measure by unstable periodic orbits in chaotic attractors. Phys. Rev. Lett. **79**, 649 (1997)
31. P. Cvitanović, G. Vattay, Entire Fredholm determinants for evaluation of semiclassical and thermodynamical spectra. Phys. Rev. Lett. **71**, 4138 (1993)
32. P. Cvitanović, Dynamical averaging in terms of periodic orbits. Phys. D. **83**, 109 (1995)
33. M.A. Zaks, D.S. Goldobin, Comment on time-averaged properties of unstable periodic orbits and chaotic orbits in ordinary differential equation systems. Phys. Rev. E. **81**, 018201 (2010)
34. Y. Nagai, Y.C. Lai, Periodic-orbit theory of the blowout bifurcation. Phys. Rev. E. **56**, 4031 (1997)

35. J. Lehnert, *Dynamics of Neural Networks with Delay*. (Master's thesis Technische Universität, Berlin, 2010)

36. E. Ott, J.C. Sommerer, Blowout bifurcations: the occurrence of riddled basins and on-off intermittency. Phys. Lett. A . **188**, 39 (1994)

37. P. Ashwin, J. Buescu, I. Stewart, From attractor to chaotic saddle: a tale of transverse instability. Nonlinearity. **9**, 703 (1996)

38. A. Englert, W. Kinzel, Y. Aviad, M. Butkovski, I. Reidler, M. Zigzag, I. Kanter, M. Rosenbluh, Zero lag synchronization of chaotic systems with time delayed couplings. Phys. Rev. Lett. **104**, 114102 (2010)

Chapter 11
Lang Kobayashi Laser Equations

Coupled semiconductor lasers will be the main application of chaos synchronization that we consider. We will therefore now introduce the dynamical laser
equations.

The equations describing a semiconductor laser with external optical feedback
were first derived by Lang and Kobayashi [1]. These Lang-Kobayashi (LK)
equations describe the laser by deterministic rate equations for the complex
electric field $\mathcal{E}$ and the number of excited carriers N in the active medium

$$\frac{d}{dt}\mathcal{E}(t) = \tfrac{1}{2}(1 + i\alpha)[\mathcal{G} - \gamma]\mathcal{E} + \kappa e^{-i\Omega_0 \tau_{ec}}\mathcal{E}(t - \tau_{ec}),$$

$$\frac{d}{dt}N(t) = \tfrac{I}{q} - \gamma_e N - \mathcal{G}|\mathcal{E}|^2 \tag{11.1}$$

with the parameters as defined in Table 11.1.

The electric field amplitude of the laser is given by $\mathcal{E}(t)e^{i\Omega_0 t}$ and is composed
of a fast carrier wave, oscillating at the solitary frequency Ω_0, and a slowly varying
envelope function $\mathcal{E}(t)$. Since the envelope function $\mathcal{E}$ is complex, it describes not
only amplitude dynamics, but also phase dynamics, i.e., shifts of the wavelength.

The feedback phase factor $e^{-i\Omega_0 \tau_{ec}}$ depends on the delay time τ_{ec} of the external
cavity. However, since Ω_0 is very large, slight changes in the delay time on
subwavelength scale change the phase drastically without changing the delayed
term of the slowly varying envelope $\mathcal{E}(t - \tau_{ec})$. It is thus useful to treat the phase
as an independent parameter φ

$$e^{i\varphi} = e^{-i\Omega_0 \tau_{ec}}.$$

The gain $\mathcal{G}$ is a function of N and $\mathcal{E}$ and for this function different forms can be
used to model the laser. A common form is a gain which is linear in N and
saturates for large $|\mathcal{E}|$

$$\mathcal{G}(\mathcal{E}, N) = g\frac{N - N_T}{1 + \epsilon|\mathcal{E}|^2};$$

see Table 11.1.

V. Flunkert, *Delay-Coupled Complex Systems*, Springer Theses,
DOI: 10.1007/978-3-642-20250-6_11, © Springer-Verlag Berlin Heidelberg 2011

Table 11.1 Typical parameters of the Lang-Kobayashi model [2]

Symbol	Quantity	Typical orders of magnitude
γ	Photon decay rate	$10^{11}\,\mathrm{s}^{-1}$
γ_e	Carrier decay rate	$10^{9}\,\mathrm{s}^{-1}$
τ_{ic}	Round-trip time in the internal cavity	$10^{-12}\,\mathrm{s}$
τ_{ec}	Round-trip time in the external cavity	$10^{-9}\,\mathrm{s}$
α	Alpha factor	4
I	Pump current	$10-100\,\mathrm{mA}$
κ	Feedback rate	$10^{11}\,\mathrm{s}^{-1}$
g	Differential gain	$10^{4}\,\mathrm{s}^{-1}$
N_T	Carrier number at transparency	10^{8}
ϵ	Gain saturation coefficient	10^{-7}
Ω_0	Solitary laser frequency	$10^{14}\,\mathrm{s}^{-1}$
β	Spontaneous emission factor	10^{-5}
q	Electron charge	$1.602 \times 10^{-19}\,\mathrm{C}$

11.1 Non-Dimensionalization

When studying dynamical systems it is convenient to bring the differential equation into a dimensionless form. This process is called non-dimensionalization [3].

Using dimensionless equations has two main advantages. It reduces the number of parameters by combining them into fewer independent constants called dimensionless groups, such as time scale ratios. Furthermore, dimensionless equations are usually better suited for numerical simulations because very large and small numbers are avoided. Often the non-dimensionalization is done by intuition but it is interesting to understand the canonical procedure.

Consider the LK equations (11.1). The general way to bring such equations into a dimensionless form is to introduce a dimensionless time s and dimensionless variables, which are functions of s and are related to the original variables by a characteristic factor carrying the proper dimensions, i.e.,

$$ s = t/t_c, \qquad \mathcal{E}(t) = \mathcal{E}_c E(t/t_c), \qquad N(t) = N_c n(t/t_c) + N_c^0. $$

The values of the characteristic factors t_c, $\mathcal{E}_c$ and N_c are to be determined. Note that we have included a constant shift N_c^0 in the transformation of the variable N. It will be chosen such that n becomes zero at the laser threshold, i.e., n is the excess carrier density or inversion. Such shifts are not necessary but can sometimes allow further simplification of the dimensionless equations. Inserting this Ansatz into (11.1) yields

$$\frac{d}{dt}\mathcal{E}(t) = \frac{\mathcal{E}_c}{t_c}\frac{d}{ds}E(s)$$

$$= \frac{1}{2}(1+i\alpha)\left[g\frac{N_c n(s) + N_c^0 - N_T}{1 + \epsilon\,|\mathcal{E}_c|^2|E(s)|^2} - \gamma\right]\mathcal{E}_c E(s)$$

$$+ \kappa e^{i\varphi}\mathcal{E}_c E(s - \tau_{ec}/t_c),$$

$$\frac{d}{dt}N(t) = \frac{N_c}{t_c}\frac{d}{ds}n(s)$$

$$= \frac{I}{q} - \gamma_e N_c^0 - \gamma_e N_c n(s) - g\frac{N_c n(s) + N_c^0 - N_T}{1 + \epsilon\,|\mathcal{E}_c|^2|E(s)|^2}|\mathcal{E}_c|^2|E(s)|^2$$

and thus

$$\frac{d}{ds}E(s) = \frac{1}{2}(1+i\alpha)\left[t_c g N_c\frac{n(s) + (N_c^0 - N_T)/N_c}{1 + |\mathcal{E}_c|^2|E(s)|^2} - \gamma t_c\right]E(s) + \kappa t_c e^{i\varphi}E(s - \tau_{ec}/t_c),$$

$$\frac{1}{\gamma_e t_c}\frac{d}{ds}n(s) = \frac{I}{q N_c \gamma_e} - \frac{N_c^0}{N_c} - n(s) - \frac{g}{\gamma_e}|\mathcal{E}_c|^2\frac{n(s) + (N_c^0 - N_T)/N_c}{1 + |\mathcal{E}_c|^2|E(s)|^2}|E(s)|^2.$$

We then choose the characteristic factors and shifts in such a way that some of the dimensionless groups become one (or zero) and the equations are simplified. The remaining dimensionless groups are then the independent parameters of the dimensionless equations and often turn out to be important characteristics such as time scale ratios. The choice which coefficients to simplify is obviously not unique in complex equations.

In our system we require the following conditions

$$1 = t_c g N_c,$$
$$1 = (N_c^0 - N_T)/N_c,$$
$$1 = \gamma t_c,$$
$$1 = \frac{g}{\gamma_e}|\mathcal{E}_c|^2.$$

This gives four equations for the three characteristic factors and for the shift of the carrier number

$$t_c = 1/\gamma, \qquad \mathcal{E}_c = \sqrt{\frac{\gamma_e}{g}}, \qquad N_c = \frac{\gamma}{g}, \qquad N_c^0 = N_T + \frac{\gamma}{g}.$$

The final dimensionless LK equations are then given by

$$\frac{d}{ds}E(s) = \frac{1}{2}(1+i\alpha)\left[\frac{1+n}{1+\mu|E|^2} - 1\right]E + Ke^{i\varphi}E(s - \tau),$$

$$T\frac{d}{ds}n(s) = p - n - \frac{1+n}{1+\mu|E|^2}|E|^2 \tag{11.2}$$

Table 11.2 Parameters of the dimensionless LK equations

Symbol	Quantity	Definition
T	Time scale parameter	γ/γ_e
τ	Delay time	$\gamma\tau_{ec}$
μ	Gain saturation coefficient	$\epsilon\gamma_e/g$
K	Feedback rate	κ/γ
p	Pump rate	$g/\gamma(\gamma_e I/q - N_T) - 1$
α	Alpha factor	
ϕ	Feedback phase	$-\Omega_0\tau_{ec}$

with the dimensionless parameters as defined in Table 11.2. Note that the dimensionless pump current p has been shifted such that it becomes zero at the laser threshold.

We will for simplicity consider the model with no gain saturation $\mu = 0$. This limit is valid when the laser is working close to threshold and the output intensity is not too large. In this case (11.2) reduce to the dimensionless LK model without gain saturation.

$$\frac{d}{ds}E(s) = \frac{1}{2}(1 + i\alpha)nE + Ke^{i\varphi}E(s - \tau),$$

$$T\frac{d}{ds}n(s) = p - n - (1 + n)|E|^2. \tag{11.3}$$

See [4] for an analysis of the gain saturation's influence on the synchronization properties.

11.2 Spontaneous Emission Noise

Important dynamical effects in semiconductor lasers are caused by stochastic forces. There are two natural sources of noise in a laser device.

On one hand there are fluctuations in the inversion, caused by shot noise, i.e., the discrete nature of the charges, as well as thermal noise. These fluctuations are usually not important for the dynamical behavior and will not be taken into account here. There are, however, exceptions where this noise cannot be neglected [5].

On the other hand there are fluctuations in the complex amplitude $\mathcal{E}$, which are caused by the spontaneous emission of photons. This noise can be modeled by a complex Gaussian white noise term $F_{\mathcal{E}}(t)$ in the LK equations (11.1)

$$\frac{d}{dt}\mathcal{E}(t) = deterministic\ terms + F_{\mathcal{E}}(t)$$

with zero mean $\langle F_{\mathcal{E}}\rangle = 0$ and the following correlations

$$\langle F_{\mathcal{E}}(t)\overline{F_{\mathcal{E}}(t')}\rangle = \beta\gamma_e N\delta(t - t'), \tag{11.4}$$

where the real and imaginary parts of $F_{\mathcal{E}}$ are independent random processes. The noise strength is given by the spontaneous emission rate

$$\mathcal{R}_{\mathrm{sp}} = \gamma_e \beta N, \tag{11.5}$$

where γ_e is the decay rate of the inversion and β is the fraction of spontaneous emission processes which contribute to the lasing mode.

   After the non-dimensionalization one obtains [6, 7] a complex Gaussian noise term F_E in (11.2) with the correlations

$$\langle F_E(t)\overline{F_E(t')}\rangle = \beta(n + n_0)\delta(t - t') \tag{11.6}$$

Here,

$$n_0 = gN_T/\gamma \tag{11.7}$$

is the carrier density at threshold in the dimensionless units. A typical value is $n_0 \approx 10$.

11.3 External Cavity Modes

The basic solutions of the LK equations are the external cavity modes (ECMs). These modes are rotating wave solutions [8, 9] with constant frequency, carrier density, and field amplitude

$$E(t) = A_* e^{i\omega_* t}, \quad n(t) = n_*.$$

Inserting this ansatz into (11.3) yields

$$0 = \frac{1}{2}n_* + K\cos(\varphi - \omega_* \tau), \tag{11.8a}$$

$$\omega_* = \frac{1}{2}\alpha n_* + K\sin(\varphi - \omega_* \tau), \tag{11.8b}$$

$$A_*^2 = \frac{p - n_*}{1 + n_*}. \tag{11.8c}$$

Eliminating n_* in the first two equations, yields a transcendental equation for the frequency ω_*

$$\omega_* = \tilde{K}\sin(\varphi - \omega_* \tau - \arctan\alpha) \tag{11.9}$$

with $\tilde{K} = K\sqrt{1 + \alpha^2}$. This transcendental equation can be solved numerically. The graphical solution is shown in Fig. 11.1(a). The solutions are born in pairs in saddle-node bifurcations. The calculated frequencies ω_* can then be inserted into (11.8) to calculate the inversion n_* and the amplitude A_*. For some solutions of

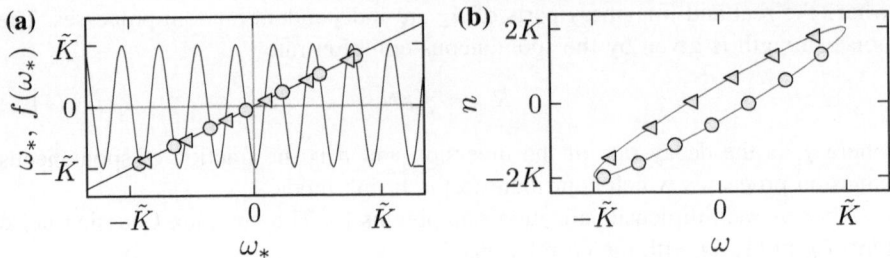

Fig. 11.1 Shown are the ECMs, where circles and triangles correspond to modes and antimodes, respectively. Panel (**a**): Solutions of the transcendental (11.9) are given by the intersection of the straight line and the curve. Panel (**b**): ECM solutions in the (ω, n)-plane. The modes and antimodes are located on the lower and upper half of the ellipse, respectively. Parameters: $T = 200$, $\alpha = 4$, $p = 1$, $\tau = 500$, $\varphi = 0$, $K = 0.01$

(11.9) the right hand side of (11.8c) may become negative. These unphysical solutions [9] are spurious and can be omitted. The solutions with a positive right hand side of (11.8c) are called physically relevant.

Figure 11.1 (b) depicts the position of the ECMs in the (ω, n)-plane, where the ECMs lie on an ellipse. To see this we introduce the curve parameter $\psi = \varphi - \omega\tau$. Inserting this ansatz into (11.8a) (11.9) yields

$$n = -2K\cos(\psi),$$

$$\omega = \tilde{K}\sin(\psi - \arctan\alpha)$$

This is the parametric representation of a tilted ellipse as depicted in Fig. 11.1b. The tilting is introduced by the additional argument $(\arctan\alpha)$ in the second equation and thus scales with the alpha factor. The mode with lowest n is called the maximum gain mode, because it is the mode with the largest amplitude. This mode is always stable [10] and coexists with a chaotic attractor for a wide range of parameters [11, 12]. Although the other ECMs are unstable, they organize the phase space and provide a skeleton for the chaotic dynamics.

There are two main chaotic operation regimes for a laser with self-feedback, which we will discuss in the following.

11.4 Low Frequency Fluctuations and Coherence Collapse

An interesting dynamical behavior of semiconductor lasers with delayed feedback are the so called low frequency fluctuations (LFFs). This regime typically occurs when the laser is pumped close to threshold and receives moderate optical feedback.

A typical time series of the laser intensity in this dynamical regime is depicted in Fig. 11.2. The intensity shows chaotic oscillations on a short time scale ($10-100\,\text{GHz}$ corresponding to $10-100\,\text{ps}$). On top of these fast oscillations there

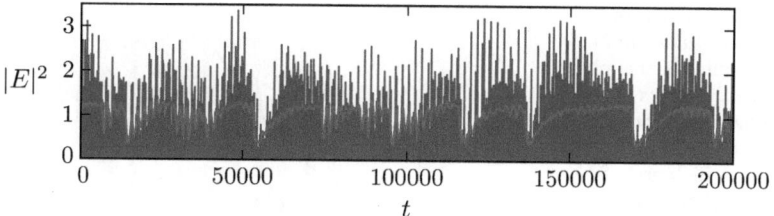

Fig. 11.2 Time series of the laser intensity in the LFF regime (*blue line*). The red curve shows a (scaled up) moving average over the intensity with a window size of 2000 time units. Parameters: $T = 200$, $p = 0.1$, $\alpha = 4$, $K = 0.1$, $\tau = 2000$, $\varphi = 0$

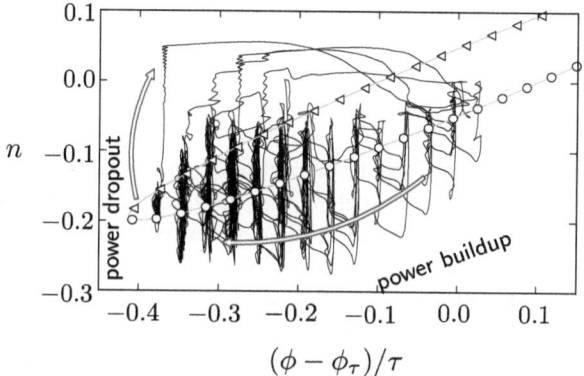

Fig. 11.3 Projection of the LFF dynamics (*black line*) onto a two dimensional phase plane. During the power buildup process, the system exhibits a chaotic itinerancy between attractor ruins with a drift towards the maximum gain mode. Before reaching the maximum gain mode, the trajectory collides with an antimode in a crisis leading to a power dropout. Modes and antimodes are depicted as circles and triangles, respectively. Parameters: $T = 200$, $\alpha = 4$, $p = 0.1$, $K = 0.1$, $\tau = 200$, $\varphi = 0$

is an occasional sudden decrease in the laser output intensity. These events are called *power dropouts* and are chaotic, too, i.e., the time between two dropouts is pseudo randomly distributed.

Typical times intervals between two dropouts reach from 10 round-trip times in the external cavity (e.g. 10 ns) to several hundred round-trip times (e.g. 1–10 ms). During the power dropouts the laser intensity is not merely reduced, but instead a sequence of picosecond pulses is generated [13, 14].

The dynamics in the LFF regime is deterministic and can be understood as follows [13–15]. In between power dropouts the intensity of the laser gradually builds up. In this phase, the dynamics is characterized by chaotic switching between attractor ruins of the laser modes. This is depicted in Fig. 11.3. The trajectory shadows heteroclinic connections between the modes. This switching

Fig. 11.4 Projection of the
CC dynamics (*black line*)
onto a two dimensional phase
plane. The dynamics is
characterized by chaotic
switching between modes and
antimodes, which are
depicted as circles and
triangles, respectively.
Parameters: $T = 200$,
$\alpha = 4$, $p = 1$, $K = 0.1$,
$\tau = 200$, $\varphi = 0$

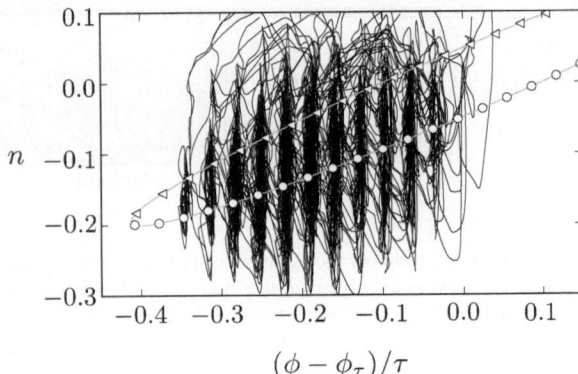

$$(\phi - \phi_\tau)/\tau$$

behavior has a general drift towards the maximum gain mode (see Chap. 3). Before reaching the maximum gain mode at the bottom of the ellipse, the trajectory collides with an antimode in a crisis and a power dropout takes place. After the dropout the trajectory gets reinjected into the labyrinth of attractor ruins and the whole process repeats itself.

For higher pump currents the laser becomes even more chaotic [16] and the LFFs disappear. This regime is called the fully developed *coherence collapse regime* [17, 11]. We will simply call it the coherence collapse (CC) regime. Here, the dynamics is characterized by chaotic itinerancy among the modes and anti-mode dynamics, i.e., collision with an antimode in a crisis. This is depicted in Fig. 11.4. In contrast to the LFF regime, however, the trajectory remains only for a short time on the mode branch of the ellipse and there is thus no distinct power buildup process and no distinct power dropout events.

In this section we have introduced and discussed the LK laser model. We will now consider two or more delay coupled lasers and study synchronization effects in these systems.

References

1. R. Lang, K. Kobayashi, External optical feedback effects on semiconductor injection laser properties. IEEE J. Quantum. Electron. **16**, 347 (1980)
2. G.P. Agrawal, N.K. Dutta, *Semiconductor Lasers*. (Van Nostrand Reinhold, New York, 1993)
3. S.H. Strogatz, *Nonlinear Dynamics and Chaos*. (Westview Press, Cambridge, 1994)
4. Hicke K. (2009) Stability of synchronized states in delay coupled lasers, Master's thesis, TU Berlin
5. M. Yousefi, D. Lenstra, G. Vemuri, Carrier inversion noise has important influence on the dynamics of a semiconductor laser. IEEE J. Sel. Top. Quantum. Electron. **10**, 955 (2004)
6. V. Flunkert, E. Schöll, Suppressing noise-induced intensity pulsations in semiconductor lasers by means of time-delayed feedback. Phys. Rev. E **76**, 066202 (2007)
7. V. Flunkert, Control of noise induced oscillations in semiconductor lasers, Master's thesis, TU Berlin (2007)

8. J. Mørk, B. Tromborg, J. Mark, Chaos in semiconductor lasers with optical feedback-Theory and experiment. IEEE J. Quantum. Electron. **28**, 93 (1992)
9. V. Rottschäfer, B. Krauskopf, The ECM-backbone of the Lang-Kobayashi equations: A geometric picture. Int. J. Bif. Chaos **17**, 1575 (2007)
10. A.M. Levine, G.H.M. van Tartwijk, D. Lenstra, T. Erneux, Diode lasers with optical feedback: Stability of the maximum gain mode. Phys. Rev. A **52**, R3436 (1995)
11. T. Heil, I. Fischer, W. Elsäßer, Coexistence of low-frequency fluctuations and stable emission on a single high-gain mode in semiconductor lasers with external optical feedback. Phys. Rev. A **58**, 2672 (1998)
12. R.L. Davidchack, Y.-C. Lai , A. Gavrielides, V. Kovanis, Dynamical origin of low frequency fluctuations in external cavity semiconductor lasers. Phys. Lett. A **267**, 350 (2000)
13. G.H.M. van Tartwijk, A.M. Levine, D. Lenstra, Sisyphus effect in semiconductor lasers with optical feedback. IEEE J. Sel. Top. Quantum. Electron. **1**, 466 (1995)
14. I. Fischer, G.H.M. van Tartwijk, A.M. Levine, W. Elsäßer, E.O Gabel, D. Lenstra, Fast pulsing and chaotic itinerancy with a drift in the coherence collapse of semiconductor lasers. Phys. Rev. Lett. **76**, 220 (1996)
15. T. Sano, Antimode dynamics and chaotic itinerancy in the coherence collapse of semiconductor lasers with optical feedback. Phys. Rev. A **50**, 2719 (1994)
16. V. Ahlers, U. Parlitz, W. Lauterborn, Hyperchaotic dynamics and synchronization of external-cavity semiconductor lasers. Phys. Rev. E **58**, 7208 (1998)
17. D. Lenstra, B. Verbeek, A. Den Boef, Coherence collapse in single-mode semiconductor lasers due to optical feedback. IEEE J. Quantum. Electron. **21**, 674 (1985)

Chapter 12
Necessary Conditions for Synchronization of Lasers

Perfect synchronization is only possible if the SM is invariant. There are other forms of *generalized synchronization* such as *phase synchronization* occurring, for instance, when the systems are non-identical, but we will restrict our analysis to perfect synchronization and a very weak form of generalized synchronization in lasers.

Consider two identical systems, which are bidirectionally coupled and have self-feedback (Fig. 12.1)

$$\frac{d}{dt}X_1 = f(X_1) + K_{11}X_1 + K_{12}X_2, \tag{12.1a}$$

$$\frac{d}{dt}X_2 = f(X_2) + K_{21}X_1 + K_{22}X_2. \tag{12.1b}$$

Here X_1 and X_2 are the state vectors of system one and two, respectively, and f is a nonlinear function. The K_{ij} are linear *coupling operators*. In the simplest case these are just matrices. However, they can also include time-shift operators, e.g.,

$$K_{12}\psi(t) = [\mathcal{T}(\tau_{12})\psi](t) = \psi(t - \tau_{12}), \tag{12.2}$$

where $\mathcal{T}(\tau)$ denotes the operator that shifts the time argument by $-\tau$. This way we can treat coupling delays in a simple way.

We consider the systems to be in (generalized) synchronization if

$$X_1 = UX_2, \tag{12.3}$$

where U is an invertible linear transformation, which leaves the dynamics of the uncoupled systems invariant, i.e., $Uf(U^{-1}X) = f(X)$. Note that this is a very weak form of generalized synchronization.

To find a synchronization condition, we introduce the symmetrized and anti-symmetrized variables $S = \frac{1}{2}(X_1 + UX_2)$ and $A = \frac{1}{2}(X_1 - UX_2)$. The original system variables can then be recovered through $X_1 = S + A$ and $X_2 = U^{-1}(S - A)$.

V. Flunkert, *Delay-Coupled Complex Systems*, Springer Theses,
DOI: 10.1007/978-3-642-20250-6_12, © Springer-Verlag Berlin Heidelberg 2011

Fig. 12.1 Schematic setup

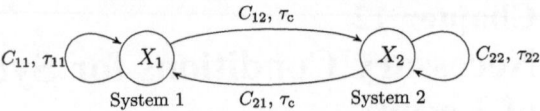

$$\text{System 1} \qquad C_{21},\, \tau_c \qquad \text{System 2}$$

If the systems are synchronized, A vanishes. Expressing the system (12.1) in S and A yields

$$\frac{d}{dt}S = \frac{1}{2}[f(S+A)+f(S-A)]$$
$$+ \frac{1}{2}\left(K_{11} + UK_{21} + K_{12}U^{-1} + UK_{22}U^{-1}\right)S$$
$$+ \frac{1}{2}\left(K_{11} + UK_{21} - K_{12}U^{-1} - UK_{22}U^{-1}\right)A, \qquad (12.4)$$

$$\frac{d}{dt}A = \frac{1}{2}[f(S+A)-f(S-A)]$$
$$+ \frac{1}{2}\left(K_{11} - UK_{21} + K_{12}U^{-1} - UK_{22}U^{-1}\right)S$$
$$+ \frac{1}{2}\left(K_{11} - UK_{21} - K_{12}U^{-1} + UK_{22}U^{-1}\right)A. \qquad (12.5)$$

In order for the SM $A = 0$ to be invariant, the coupling term in front of S in (12.5) has to vanish

$$0 = K_{11} - UK_{21} + K_{12}U^{-1} - UK_{22}U^{-1}. \qquad (12.6)$$

This can also be interpreted as an equation for the transformation U. If there exists a solution U solving (12.6), which is not always the case, then the systems can synchronize according to (12.3). Whether the synchronization is stable, is another question which we answered for large delays in Chap. 10.

If (12.6) is satisfied we can add the right hand side to each coupling term in (12.4) and (12.5) to simplify the equations for the synchronized dynamics. Linearizing the equations with respect to A around the synchronized state $A = 0$ then gives

$$\frac{d}{dt}S = f(S) + \left(K_{11} + K_{12}U^{-1}\right)S + \left(K_{11} - UK_{22}U^{-1}\right)\delta A, \qquad (12.7)$$

$$\frac{d}{dt}\delta A = Df(S)\,\delta A + \left(K_{11} - UK_{21}\right)\delta A. \qquad (12.8)$$

Let us now discuss the δA term in (12.7). We want to discuss the stability of the SM, i.e., whether infinitesimal perturbations to $A = 0$ grow or decay. Since we consider the case of chaos synchronization, the dynamics of S is chaotic. Small perturbations caused by the δA term in (12.7) will change the specific realization of the chaotic S-trajectory but not the statistical properties. In particular if the perturbations are small enough the shadowing lemma [1] ensures the existence of a shadowing trajectory in the unperturbed system, i.e., a trajectory, which follows the trajectory of the perturbed system arbitrarily close and for an arbitrarily long time. Thus the δA term does not influence stability of the SM.

With this discussion the equations determining the synchronized dynamics and the transverse stability are given by

$$\frac{d}{dt}S = f(S) + \left(K_{11} + K_{12}U^{-1}\right) S, \tag{12.9}$$

$$\frac{d}{dt}\delta A = Df(S)\,\delta A + (K_{11} - UK_{21})\,\delta A. \tag{12.10}$$

Note that although the transversal Lyapunov exponent (TLE) will usually not be influenced by the δA term, the bubbling dynamics may be changed [2].

From these equations we can make the following general observation. If

$$K_{12}U^{-1} = -UK_{21},$$

then the variational equation for the parallel LE and the TLE are the same, i.e.,

synchronized solution chaotic $\Leftrightarrow$ synchronization unstable.

As we saw in Chap. 10 this relation between chaoticity and synchronization applies to many other networks (type (B) and (C) with $|\gamma_{max}| \leq |\sigma|$) in the case of large delay. Similar results are known for other types of coupling [3]. It is usually more difficult to achieve chaos synchronization than non-chaotic synchronization.

12.1 Coupling Delays

Let us investigate the role of the coupling delays. Writing the coupling operators K_{jl} as a time shift $T(\tau_{jl})$ and a coupling matrix C_{jl} we find

$$K_{jl} = C_{jl}T(\tau_{jl}) \quad \text{and} \quad U = C_u T(\tilde{\tau}_u),$$

and (12.6) becomes

$$\begin{aligned} 0 = {} & C_{11}T(\tau_{11}) - C_u C_{21}T(\tau_{21} + \tilde{\tau}_u) \\ & + C_{12}C_u^{-1}T(\tau_{12} - \tilde{\tau}_u) - C_u C_{22}C_u^{-1}T(\tau_{22}). \end{aligned} \tag{12.11}$$

We can simplify [4] the equation a little by introducing the new parameter τ_u and choosing

$$\tilde{\tau}_u = \tau_u + (\tau_{12} - \tau_{21})/2, \tag{12.12}$$

i.e., compensating a shift caused by $\tau_{12} \neq \tau_{21}$. This gives two identical delay terms $\tau_c = (\tau_{12} + \tau_{21})/2$ in (12.11)

$$0 = C_{11}\mathcal{T}(\tau_{11}) - C_u C_{21}\mathcal{T}(\tau_c + \tau_u)$$
$$+ C_{12}C_u^{-1}\mathcal{T}(\tau_c - \tau_u) - C_u C_{22}C_u^{-1}\mathcal{T}(\tau_{22}). \tag{12.13}$$

And it follows that without loss of generality the coupling delays τ_{12} and τ_{21} can be chosen equal.

The right hand side of (12.13) can only become zero if the coefficient in front of each *independent* time shift operator vanishes, i.e., each term with non-zero C_{jl} cancels with other terms, and this is only possible if the time shifts are equal.

Consider, for example, the case where two coupling coefficients are zero. In order for the remaining two terms to cancel the delay terms have to be equal and the coefficients have to add up to zero.

The different ways the terms in (12.13) can cancel lead to different coupling schemes and synchronization conditions. Before we give a comprehensive overview of these coupling schemes, we will explain the connection to lasers. In the general case the equations for two delay-coupled lasers are given by (see Chap. 11)

$$\frac{d}{dt}E_j = \frac{1}{2}(1 + i\alpha)n_j E_j + \kappa_{jj}e^{i\phi_{jj}} E_j(t - \tau_{jj}) + \kappa_{jl}e^{i\phi_{jl}} E_l(t - \tau_{jl}),$$
$$T\frac{d}{dt}n_j = p - n_j - (1 + n_j)|E_j|^2$$

with $j = 1, 2$ and $l = 3 - j$. Here, κ_{jl} are the positive feedback gains and the exponential terms account for shifts in the optical phase. The phases ϕ_{jj} and ϕ_{jl} depend on subwavelength tuning of the cross and self-feedback delays. However, the phases can be considered as being independent of the delay times as discussed in Chap. 11. For all-optical coupling we can neglect the carriers in (12.13) and only have to take the coupling terms in the field equation into account.

The only transformations C_u, which leave the laser equations invariant are phase rotations, due to the S^1-symmetry of the system. In particular any linear transformation that changes the intensity or carrier density does not leave the equations invariant. Thus we make the ansatz

$$C_u = e^{i\tilde{\phi}_u},$$

which corresponds to the lasers having a constant phase-shift $\tilde{\phi}_u$ in the electric fields (in addition to the time lag of τ_u). The synchronization condition (12.13) then reads in the general case

$$0 = \kappa_{11}e^{i\phi_{11}}\mathcal{T}(\tau_{11}) - \kappa_{21}e^{i(\phi_{21}+\tilde{\phi}_u)}\mathcal{T}(\tau_c + \tau_u)$$
$$+ \kappa_{12}e^{i(\phi_{12}-\tilde{\phi}_u)}\mathcal{T}(\tau_c - \tau_u) - \kappa_{22}e^{i\phi_{22}}\mathcal{T}(\tau_{22}). \tag{12.14}$$

Let us now come to the general problem of solving (12.13) or for lasers (12.14). The different ways the terms can cancel correspond to different coupling schemes.

Table 12.1 shows all possible ways for the terms to cancel, leaving out redundant cases, which correspond to relabeling $1 \longleftrightarrow 2$. In the label column the roman number denotes how many links are present, and the letter enumerates these cases. The non-zero coefficients in the second column correspond to the links present in the network motif [5]. Then, in order for the remaining terms to cancel, the discussed conditions on the non-vanishing coefficients and possibly the coupling delays arise, and are given in columns 4 and 5. The last two columns give the equations of motion within the SM and the time lag τ_u in the synchronized state between the two systems. In the cases IVa–IVc of four non-zero coefficients, respectively two terms in (12.14) have the same delay time and cancel each other and are depicted with the line style (solid or dashed) in Table 12.1. In the case IVd all terms have the same delay time and thus the sum of all coefficients is zero.

We are interested in chaos synchronization and we can distinguish different situations. An individual system may already be chaotic without any feedback or it may be chaotic only with self-feedback, as is the case in semiconductor lasers. For the latter type of systems the coupling scheme IIa and IVa do not have a chaotic solution, since there is no feedback term in the equation of motion in the synchronization manifold. Furthermore, for those systems, which are chaotic without self-feedback synchronization is unstable in these coupling schemes for large delay times (see [6] and Chap. 10). Therefore, such coupling schemes are not interesting for chaos synchronization and we will not discuss the two coupling schemes IIa and IVa.

The coupling scheme IId is not interesting for chaos synchronization, either, since the two systems are completely uncoupled. In this case if each system evolves on a PO, the "synchronized state" is marginally stable but if the dynamics is chaotic the synchronized state is of course unstable.

We will now discuss the remaining coupling schemes in Table 12.1 and the relation to lasers. We will concentrate on the necessary synchronization conditions for lasers, which lead to coupling phase conditions. In general the coupling can be unidirectional, corresponding to a master-slave setup, or bidirectional, where both lasers receive input from each other. Additionally, we can distinguish between open-loop setups and closed-loop setups. These terms are coined by control theory. Closed-loop corresponds to the lasers receiving self-feedback, whereas in open-loop setups the lasers do not receive self-feedback.

IIb This case is the classical master slave configuration for chaos communication with lasers. It is also referred to as open-loop master slave configuration [7], since the receiver has no self-feedback. The coupling condition for lasers is in this case given by

$$0 = \kappa_{11} e^{i\phi_{11}} - \kappa_{21} e^{i(\phi_{21}+\tilde{\phi}_u)}.$$

Thus, the only condition that needs to be satisfied is $\kappa_{11} = \kappa_{21}$ (at least approximately). The phase shift $\tilde{\phi}_u$ between the lasers can then compensate

Table 12.1 Classification of the synchronization properties for different coupling schemes of two delay coupled systems

Label	Non-zero coefficients	Topology	Coupling conditions	Delay conditions	Dynamics in the synchronized state	Time lag
IIa	C_{22}, C_{21}		$0 = C_{21} + C_{22}C_u^{-1}$	—	$\dot{S} = f(S) + 0$	$\tau_u = \tau_{22} - \tau_c$
IIb	C_{11}, C_{21}		$0 = C_{11} - C_u C_{21}$	—	$\dot{S} = f(S) + C_{11}S_{\tau_{11}}$	$\tau_u = \tau_{11} - \tau_c$
IIc	C_{21}, C_{12}		$0 = C_{12}C_u^{-1} - C_u C_{21}$	—	$\dot{S} = f(S) + C_u C_{21}S_{\tau_c}$	$\tau_u = 0$
IId	C_{11}, C_{22}		$0 = C_{11} - C_u C_{22}C_u^{-1}$	$\tau_{11} = \tau_{22}$	$\dot{S} = f(S) + C_{11}S_{\tau_{11}}$	τ_u arbitrary
IIIa	C_{21}, C_{12}, C_{22}		$0 = C_{12}C_u^{-1} - C_u C_{21} - C_u C_{22}C_u^{-1}$	$\tau_{22} = \tau_c$	$\dot{S} = f(S) + C_{12}C_u^{-1}S_{\tau_c}$	$\tau_u = 0$
IIIb	C_{22}, C_{11}, C_{21}		$0 = C_{11} - C_u C_{21} - C_u C_{22}C_u^{-1}$	$\tau_{11} = \tau_{22}$	$\dot{S} = f(S) + C_{11}S_{\tau_{11}}$	$\tau_u = \tau_{22} - \tau_c$
IVa	all		$0 = C_{11} + C_{12}C_u^{-1},\ 0 = C_{21} + C_{22}C_u^{-1}$	$\tau_{11} + \tau_{22} = 2\tau_c$	$\dot{S} = f(S) + 0$	$\tau_u = \tau_{22} - \tau_c$
IVb	all		$0 = C_{11} - C_u C_{21},\ 0 = C_{12} - C_u C_{22}$	$\tau_{11} + \tau_{22} = 2\tau_c$	$\dot{S} = f(S) + C_{11}S_{\tau_{11}} + C_{12}C_u^{-1}S_{\tau_{22}}$	$\tau_u = \tau_c - \tau_{22}$
IVc	all		$0 = C_{11} - C_u C_{22}C_u^{-1},\ 0 = C_{12}C_u^{-1} - C_u C_{21}$	$\tau_{11} = \tau_{22}$	$\dot{S} = f(S) + C_{11}S_{\tau_{11}} + C_u C_{21}S_{\tau_c}$	$\tau_u = 0$
IVd	all		$0 = C_{11} - C_u C_{21} + C_{12}C_u^{-1} - C_u C_{22}C_u^{-1}$	$\tau_{11} = \tau_{22} = \tau_c$	$\dot{S} = f(S) + (C_{11} + C_{12}C_u^{-1})S_{\tau_{11}}$	$\tau_u = 0$

any choice of the coupling phases ϕ_{11} and ϕ_{21}. Thus we do not observe any sensitivity to the phases.

IIc This coupling scheme corresponds to the well known configuration of two bidirectionally coupled systems without self-feedback. As we saw in Chap. 10 zero-lag chaos synchronization is unstable for large delay. For chaotic lasers coupled in this fashion a different type of synchronized called *leader-laggard synchronization* has been observed [8]. Here one laser lags behind the other with a time shift of τ_c. The role of the leader and the laggard then switch chaotically. Note that we do not find an exact solution of one laser leading the other, thus this type of synchronization is not an exact synchronization but only an approximate.

IIIa In this situation all delays are the same, the coupling is bidirectional and one of the elements has a self-feedback. The necessary coupling conditions for this case are very similar to the case IIIb, which we will discuss in detail below.

IIIb Two unidirectionally coupled systems with self-feedback have been studied in laser systems in different contexts.

Depending on the values of $\tau_{22} = \tau_{11}$ and τ_c the time lag

$$\tau_u = \tau_{22} - \tau_c$$

between the lasers may be positive or negative. Since the states of the two laser are related by (see (12.3))

$$X_1(t) \propto X_2(t - \tau_u),$$

for positive τ_u laser *2* is ahead of laser *1* although the coupling is from laser 1 to laser 2 [9]. This behavior is known as *anticipated synchronization* and generally occurs in delayed systems if the coupling delay is smaller than the self-feedback delay [10]. In fact, for this coupling scheme that we consider the synchronization properties are independent of the delay time τ_c, since there is no link back to laser 1.

The importance of the phases in this coupling scheme has been recognized in [11]. We will discuss this in detail in Sect. 12.2.1.

IVb This is the case of two bidirectionally coupled systems with self-feedback, where the two self-feedback delays differ $\tau_{11} \neq \tau_{22}$ but sum up to the round-trip time between the laser $\tau_{11} + \tau_{22} = 2\tau_c$.

It corresponds to the two systems being coupled via a passive relay with a delay miss-match [12, 13], e.g., two laser coupled via a semitransparent mirror positioned asymmetrically between the lasers. The dynamics in this case is quite complicated. In the synchronized state the system behaves like a single laser with two feedback delays. One can study the transverse stability of the modes in the SM, similar to the analysis that we will perform in Chap. 13.

In the present work, we will restrict our analysis to the role of the phases for this case (see Sect. 12.2.3).

IVc   In this case the lasers are coupled such that the two self-feedback delays are
      equal $\tau_{11} = \tau_{22}$ but do not match the coupling delay τ_c. This case is similar to
      the case IVd (see Sect. 12.2.3) and will not be discussed separately.
IVd   For this case all delays are equal and all four coefficients in (12.14) cancel
      collectively. We will discuss this case in detail in Sect. 12.2.2 for lasers

12.2 Role of the Phases

12.2.1 Case IIIb

In Ref. [11] it was shown that depending on the (relative) feedback phases the
synchronization behavior ranges from perfect synchronization to an almost
uncorrelated state. This result can be interpreted with our necessary coupling
condition. Using the coupling coefficients

$$C_{11} = \kappa_{11}e^{i\phi_{11}}, \quad C_{22} = \kappa_{22}e^{i\phi_{22}}, \quad \text{and} \quad C_{21} = \kappa_{21}e^{i\phi_{21}},$$

the coupling condition from Table 12.1 becomes

$$0 = \kappa_{11}e^{i\phi_{11}} - \kappa_{21}e^{i(\phi_{21}+\tilde{\phi}_u)} - \kappa_{22}e^{i\phi_{22}}.$$

In order for these vectors in the complex plane to cancel, the following equation
needs to be satisfied

$$\kappa_{21} = \left|\kappa_{11} - \kappa_{22}e^{i\Phi_{rel}}\right| \tag{12.15}$$

with $\Phi_{rel} = \phi_{22} - \phi_{11}$. The relative phase $\tilde{\phi}_u$ between the laser fields can then be
selected by the system accordingly.

Figure 12.2(a) depicts the square difference of the absolute value of equation's
(12.15) left hand side and right hand side

$$V := \left(\kappa_{21} - \left|\kappa_{11} - \kappa_{22}e^{i\Phi_{rel}}\right|\right)^2$$

as a function of Φ_{rel}. This is a measure for how much the necessary synchronization condition (12.15) is violated. Panel (b) of this figure depicts the correlation
coefficient between the intensities of laser one and two as measured by Peil et al.
in Ref. [11]. Our theory predicts an invariant SM and thus maximum correlation
for $\Phi_{rel} \approx 0.22\pi$ and $\Phi_{rel} \approx 1.78\pi$. From the experimental (black squares) and
numerical data (triangles) in the right panel it is not completely clear whether
maxima are present at these Φ_{rel} values. However, the numerical data seem to
suggest a small local minimum of the correlation at $\Phi_{rel} = 0 \triangleq 2\pi$. Other feedback

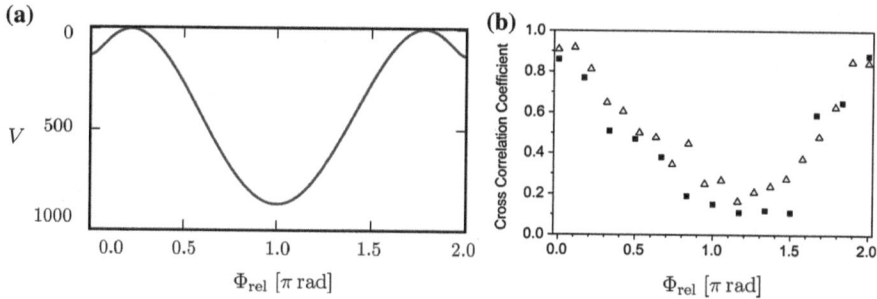

Fig. 12.2 Panel (**a**): Square difference between the absolute values of the left hand side and the right hand side of (12.15) vs Φ_{rel}, measuring how much the synchronization condition is violated. Coupling coefficients as in [11] $\kappa_{11} = 25\,\text{ns}^{-1}$, $\kappa_{22} = 20.5\,\text{ns}^{-1}$, $\kappa_{21} = 16\,\text{ns}^{-1}$. Panel (**b**): Correlation coefficient of the first and second lasers intensities vs. Φ_{rel}. Squares and triangles mark experimental and numerical data, respectively (Figure in panel (*b*) from [11] courtesy I. Fischer[1]

gains should shift the maxima and minima in the correlation to other values of Φ_{rel} and it would be interesting to investigate this further.

12.2.2 Case IVd

We now consider the coupling scheme IVd of all coupling delays being equal $\tau = \tau_{jj} = \tau_{jl}$ for lasers. The necessary synchronization condition from Table 12.1 becomes for the laser case

$$0 = \kappa_{11}e^{i\phi_{11}} - \kappa_{21}e^{i(\phi_{21}+\tilde{\phi}_{\text{u}})} + \kappa_{12}e^{i(\phi_{12}-\tilde{\phi}_{\text{u}})} - \kappa_{22}e^{i\phi_{22}}. \qquad (12.16)$$

We are trying to find explicit conditions on the coupling strengths κ_{jl} and phases ϕ_{jl}, such that a phase relation $\tilde{\phi}_{\text{u}}$ between the lasers exists, which solves (12.16). Rotating all phases by $\theta := -(\phi_{12} + \phi_{21})/2$ and using

$$\phi_{\text{u}} := \tilde{\phi}_{\text{u}} + \frac{\phi_{21} - \phi_{12}}{2} \qquad (12.17)$$

we obtain the simplified equation

[1]   Reprinted figure with permission from Michael Peil, Tilmann Heil, Ingo Fischer and Wolfgang Elsäßer, Phys. Rev. Lett. **88**, 174101 (2002). Copyright 2010 by the American Physical Society.)

Fig. 12.3 Geometric
visualization of the
synchronization condition.
The black rods can rotate
around the joints. If the end
of the second rod lies on the
ellipse, the SM is invariant.
The existence of a solution
depends on the rod lengths
and the lengths of the
ellipse's semi axis

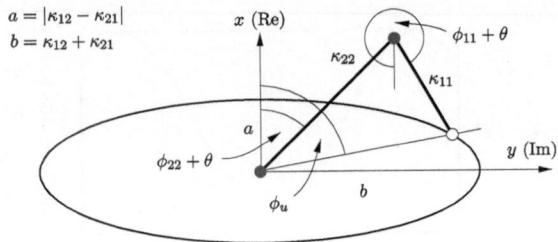

$$\kappa_{22}e^{i(\phi_{22}+\theta)} - \kappa_{11}e^{i(\phi_{11}+\theta)} = \kappa_{12}e^{-i\phi_u} - \kappa_{21}e^{i\phi_u}. \qquad (12.18)$$

Note how similar this transformation is to (12.12).

For varying ϕ_u the terms on the right hand side of (12.18) describe an ellipse in the complex plane with semi-minor axis $a = |\kappa_{12} - \kappa_{21}|$ oriented along the real axis and semi-major axis $b = |\kappa_{12} + \kappa_{21}|$ oriented along the imaginary axis.

For the equation to have a solution the real and imaginary part of the left hand side have to lie on this ellipse. Thus

$$x := \kappa_{22}\cos(\phi_{22} + \theta) - \kappa_{11}\cos(\phi_{11} + \theta) \qquad (12.19a)$$

$$y := \kappa_{22}\sin(\phi_{22} + \theta) - \kappa_{11}\sin(\phi_{11} + \theta) \qquad (12.19b)$$

have to obey

$$b^2 x^2 + a^2 y^2 = a^2 b^2. \qquad (12.20)$$

Note that we use this form of the ellipse equation to include the degenerate case $\kappa_{12} = \kappa_{21}$, i.e., $a = 0$ for which we have to explicitly consider the allowed ranges of x and y

$$x^2 \le a^2 \text{ and } y^2 \le b^2. \qquad (12.21)$$

Equation (12.20) is the final condition, which has to be fulfilled in order for the SM to be invariant. It is thus a necessary condition for synchronization. It involves all coupling strengths and coupling phases but not the relative phase shift ϕ_u between the lasers. This relative phase shift can be found by solving (12.18) for ϕ_u

$$\tan\phi_u = \frac{\kappa_{21} - \kappa_{12}}{\kappa_{21} + \kappa_{12}}\frac{y}{x}, \qquad (12.22)$$

which then gives the actual phase shift $\tilde{\phi}_u$ of the lasers via (12.17).

The problem of solving (12.20) can be visualized as follows. Consider the geometric situation sketched in Fig. 12.3. Two rods are connected by a joint with

Fig. 12.4 Solutions of the
phase condition. Solution of
the phase condition (12.20)
for different values of
$b = |\kappa_{12} + \kappa_{21}|$. Other
parameters: $\kappa_{11} = 0.15$,
$\kappa_{22} = 0.25$,
$a = |\kappa_{12} - \kappa_{21}| = 0.2$

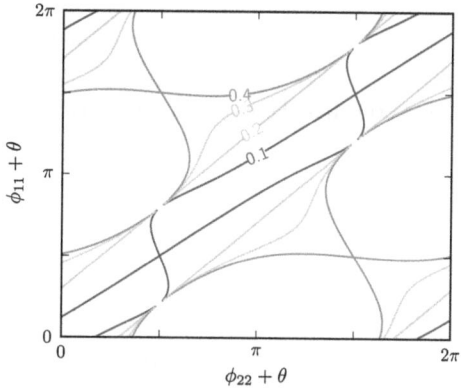

each other and to the origin. The end of the second rod can slide on the ellipse. The question is for which values of the semi-major axis a and b and rod lengths κ_{11} and κ_{22} do solutions exist and if solutions exist what is the relation between the angles $\phi_{11} + \theta$ and $\phi_{22} + \theta$.

The connection to this geometrical problem is obvious. Equations (12.19) describe points, which can be reached by adding two vectors of lengths κ_{22} and κ_{11}. According to (12.20) this sum has to lie on the ellipse. Note that it does not matter which of the two rods is connected to the origin, since vector addition is commutative.

The only free phase parameters are $\phi_{11} + \theta$ and $\phi_{22} + \theta$. A change of the other phases, will only influences the constant phase shift ϕ_u between the two lasers. Figure 12.4 depicts solutions of (12.20) for different values of the coupling strengths.

In order for (12.20) to have a solution, the rods have to be able to reach the ellipse, i.e., the sum of the rod lengths $\kappa_{11} + \kappa_{22}$ has to be larger or equal to the semi-minor axis $a = |\kappa_{12} - \kappa_{21}|$. Similarly, the absolute value of the rod lengths difference $|\kappa_{11} - \kappa_{22}|$ has to be smaller or equal to the large semi-major axis $b = \kappa_{12} + \kappa_{21}$. This gives two conditions for the existence of a solution

$$(\kappa_{11} + \kappa_{22})^2 \geq (\kappa_{12} - \kappa_{21})^2, \quad \text{and} \tag{12.23a}$$

$$(\kappa_{11} - \kappa_{22})^2 \leq (\kappa_{12} + \kappa_{21})^2. \tag{12.23b}$$

If and only if the coupling strengths fulfill (12.23), there is a combination of phases such that condition (12.20) is satisfied.

We will now consider the case $\kappa_{12} = \kappa_{21}$. This case is important for applications, because an optical face-to-face setup with a partially-transparent mirror will always have $\kappa_{12} \approx \kappa_{21}$ the other feedback strengths can however be chosen differently, due to different transparency and reflectivity of the mirror or grey filters in the optical beam. For $a = 0$ the ellipse becomes a line along the y-axis stretching from $-b$ to b. Thus (12.20) reduces to $x = 0$ with the side condition $y^2 \leq b^2$. We assume without loss of generality $\kappa_{11} \leq \kappa_{22}$ and find

$$\phi_{22} + \theta = \pm \arccos\left(\frac{\kappa_{11}}{\kappa_{22}} \cos(\phi_{11} + \theta)\right) \tag{12.24}$$

If we consider the even simpler case $\kappa_{11} = \kappa_{22}$, i.e., the two lasers have the same self-feedback strengths (12.24) simplifies to

$$\phi_{22} + \theta = \pm(\phi_{11} + \theta) \tag{12.25}$$

and we have two solutions

$$\phi_{11} = \phi_{22}, \quad \text{and} \tag{12.26a}$$

$$\phi_{11} + \phi_{22} = \phi_{12} + \phi_{21}. \tag{12.26b}$$

For the first solution arbitrary values of ϕ_{12} and ϕ_{21} are allowed.

Note that we still have the side condition $y^2 \le b^2$. This is always fulfilled for the solution (12.26a) since $y = 0$ in this case. For solution (12.26b) the restriction $y^2 \le b^2$ is relevant if $\kappa_{21} = \kappa_{12} < \kappa_{11} = \kappa_{22}$ and in this case only phases, which obey

$$\sin^2(\phi_{22} + \theta) < \kappa_{12}^2/\kappa_{11}^2$$

allow a synchronized solution.

In either case (12.22) cannot be used to calculate the phase shift ϕ_u between the lasers, since $0 = \kappa_{12} - \kappa_{21}$ as well as $x = 0$. Going back to (12.18), we find the following phase shifts for the two cases

$$\phi_{11} = \phi_{22}, \rightarrow \quad \phi_u = 0, \quad \text{or} \quad \phi_u = \pi,$$
$$\phi_{11} + \phi_{22} = \phi_{12} + \phi_{21} \rightarrow \quad \phi_u = \phi_{22} - \phi_{21}. \tag{12.27}$$

Note that this phase condition is naturally fulfilled for a perfect experimental setup with a semitransparent mirror (at any position) in between the lasers, where the optical path lengths obey

$$\tau_{11} + \tau_{22} = \tau_{21} + \tau_{12}$$

on a subwavelength scale.

To confirm (12.26) we simulated the laser equations for fixed $\phi_{11} = 0$ varying the other three phases between 0 and 2π with a mesh size of $\pi/4$ resulting in $8 \times 8 \times 8 = 512$ points in the $(\phi_{12}, \phi_{21}, \phi_{22})$ parameter space. Figure 12.5 depicts the result. The size and brightness of the glyphs display the magnitude of the correlation coefficient

$$r = \frac{\langle [I_1 - \langle I_1 \rangle][I_2 - \langle I_2 \rangle] \rangle}{\Delta I_1^2 \, \Delta I_2^2}$$

Fig. 12.5 Correlation
coefficient r for different
values of the phases
ϕ_{12}, ϕ_{21}, ϕ_{22}. The size and
brightness of the glyphs
depict the magnitude of the
correlation coefficient. The
correlation is largest on the
blue planes where the phase
condition (12.27) is fulfilled.
Parameters: $\phi_{11} = 0$, $\alpha = 4$,
$T = 200$, $p = 5$,
$K = 0.04$, $\tau = 1000$

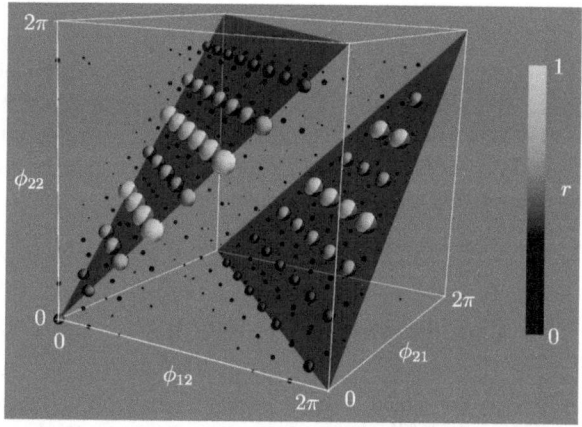

of the intensities where ΔI_1 and ΔI_2 denote the standard deviations of the intensities I_1 and I_2, respectively. The blue planes show the solution of (12.27). The correlation is only strong if the phase condition is satisfied. Even for satisfied phase condition, the synchronized solution may have different stability properties depending on the values of the phases. This results in different values of the cross correlation within the blue planes in Fig. 12.5.

12.2.3 Case IVb

Let us now consider the coupling scheme IVb, corresponding to two unequal self-feedback delays $\tau_{11} \neq \tau_{22}$, which add up to twice the coupling delay $\tau_{11} + \tau_{22} = 2\tau_c$.

As discussed above (see also Chap. 13), for lasers this case can be realized experimentally with a semitransparent mirror in between the lasers positioned asymmetrically, leading to a delay mismatch [12, 13].

From the corresponding coupling conditions in Table 12.1 we find the following two conditions on the coupling strengths and phases

$$0 = \kappa_{11}e^{i\phi_{11}} - \kappa_{21}e^{i(\phi_{21}+\phi_u)}, \qquad (12.28a)$$

$$0 = \kappa_{12}e^{i\phi_{12}} - \kappa_{22}e^{i(\phi_{22}+\phi_u)}. \qquad (12.28b)$$

These equations can only be satisfied for

$$\kappa_{11} = \kappa_{21}, \quad \text{and} \quad \kappa_{12} = \kappa_{22}. \qquad (12.29)$$

Assuming the coupling strengths obey (12.29), we can eliminate ϕ_u form (12.28) and obtain the same phase condition as in the last section

$$\phi_{11} + \phi_{22} = \phi_{12} + \phi_{21}.$$

Although this case is mathematically easier than the case IVd above, it will be more difficult to find synchronization, experimentally. Here, we have to meet three conditions (two on the coupling strengths, and one on the phases) in contrast to the case IVd in the last section, where only two conditions (one on the coupling strengths and one on the phases) needed to be met.

12.3 Conclusion and Outlook

We have seen in this section that the coupling phases play a crucial role for the synchronizability of lasers coupled all-optically in network motifs.

The necessary phase conditions are essentially interference conditions. As such, they arise as soon as one of the lasers has more than one input. Then the interference condition demands that the input signals interfere such that each laser has the same input signal, relative to its own phase.

These interference conditions pose a great problem for experiments. The phases are sensitive to changes of the optical path lengths on the subwavelength scale. Although the phases can be controlled via the current through a passive phase section (see, for instance, [14]), it seems unreasonable to control the phases in a larger network such that all phase conditions are matched. For the synchronization of larger networks optoelectronic coupling, which is insensitive to phases, seems therefore more promising.

References

1. C. Grebogi, S.M. Hammel, J.A. Yorke, T. Sauer, Shadowing of physical trajectories in chaotic dynamics: Containment and refinement. Phys. Rev. Lett. **65**, 1527 (1990)
2. P. Ashwin, E. Covas, R. Tavakol, Transverse instability for non-normal parameters. Nonlinearity **12**, 563 (1999)
3. A.S. Pikovsky, M.G. Rosenblum, J. Kurths, Synchronization, A Universal Concept in Nonlinear Sciences. (Cambridge University Press, Cambridge, 2001)
4. A. Panchuk, M.A. Dahlem, E. Schöll, Regular spiking in asymmetrically delay-coupled FitzHugh-Nagumo systems. Proceedings NDES 09. (2009), pp. 177–179 arXiv:0911.2071v1
5. O. D'Huys, R. Vicente, T. Erneux, J. Danckaert, I. Fischer, Synchronization properties of network motifs: Influence of coupling delay and symmetry. Chaos **18**, 037116 (2008)
6. W. Kinzel, A. Englert, G. Reents, M. Zigzag, I. Kanter, Synchronization of networks of chaotic units with time-delayed couplings. Phys. Rev.E **79**, 056207 (2009)
7. R. Vicente, T. Pérez, C.R. Mirasso, Open-versus closed-loop performance of synchronized chaotic external-cavity semiconductor lasers. IEEE J. Quantum Electron. **38**, 1197 (2002)
8. J. Mulet, C.R. Mirasso, T. Heil, I. Fischer, Synchronization scenario of two distant mutually coupled semiconductor lasers. J. Opt. B **6**, 97 (2004)

9. C. Masoller, Anticipation in the synchronization of chaotic semiconductor lasers with optical feedback. Phys. Rev. Lett. **86**, 2782 (2001)
10. H.U. Voss, Anticipating chaotic synchronization. Phys. Rev. E **61**, 5115 (2000)
11. M. Peil, T. Heil, I. Fischer, W. Elsäßer, Synchronization of chaotic semiconductor laser systems: A vectorial coupling-dependent scenario. Phys. Rev. Lett. **88**, 174101 (2002)
12. K. Hicke, Stability of synchronized states in delay coupled lasers, Master's thesis. TU Berlin, (2009)
13. K. Hicke, O. D'Huys, V. Flunkert, E. Schöll, J. Danckaert, I. Fischer, Mismatch and synchronization: Influence of asymmetries in systems of two delay-coupled lasers. Phys. Rev. E, (2011) submitted
14. H.J. Wünsche, O. Brox, M. Radziunas, F. Henneberger, Excitability of a semiconductor laser by a two-mode homoclinic bifurcation. Phys. Rev. Lett. **88**, 023901 (2002)

Chapter 13
Bubbling

The stability of a synchronized state is determined by the largest transversal Lyapunov exponent (TLE) arising from the particular dynamics in the SM and the variational equation associated with transverse perturbations, as we have discussed in Chap. 10.

If the largest TLE is negative, the synchronized state is linearly stable. There is, however, a nonlinear effect, which can render the synchronization unstable in the presence of noise or parameter mismatch. This effect is called bubbling [1] or riddling [2] and is associated to transversely unstable invariant sets in the attractor.

Bubbling can occur when a dynamical system has an invariant manifold and embedded in this manifold is a chaotic attractor. The most common situation with this requirement is chaos synchronization of two coupled systems. In this case the synchronization manifold is invariant and the synchronized chaotic dynamics is restricted to the synchronization manifold. In the following we will discuss this situation for delay coupled lasers.

The transverse stability of an orbit in the manifold, i.e., the stability in the direction perpendicular to the manifold, is determined by the transversal Lyapunov exponent of the orbit. It is important to note that each invariant set in the manifold, i.e., each FP, PO or chaotic orbit, has a distinct transversal Lyapunov spectrum and therefore distinct stability properties. However, if the system is chaotic almost all initial conditions in the SM will lie on the chaotic attractor and thus give rise to the same TLE. The term TLE is then often used for this exponent, arising from almost any initial conditions.

Within any chaotic attractor there are always infinitely many UPOs and the chaotic behavior can in fact be characterized through all POs in the attractor [3]. When such an UPO in the attractor becomes transversely unstable while the chaotic attractor itself is still transversely stable, the trajectory can be pushed towards the transversely unstable orbit even by arbitrarily small noise and then leave the invariant manifold. Thus the transversely unstable orbits provide escape routes from the attractor. A cartoon of this situation is depicted in Fig. 13.1. Almost all trajectories in the SM are transversely stable and only some trajectories (with measure zero) marked by the red points are transversely unstable.

V. Flunkert, *Delay-Coupled Complex Systems*, Springer Theses,
DOI: 10.1007/978-3-642-20250-6_13, © Springer-Verlag Berlin Heidelberg 2011

Fig. 13.1 The transversely
unstable orbits (*red dots*)
provide escape routes from
the transversely stable chaotic
attractor

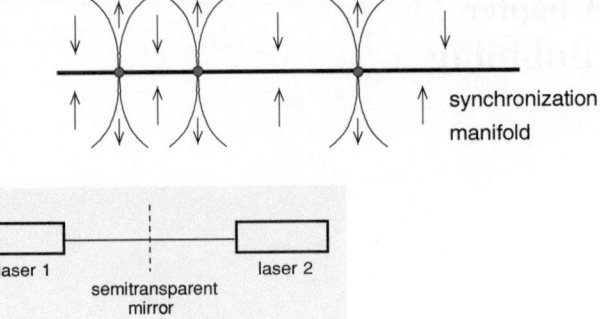

Fig. 13.2 Schematic setup of two lasers delay coupled via a semitransparent mirror. Each lasers receives self-feedback and input from the other laser. The coupling and self-feedback strengths are determined by the transmittance and reflectivity of the mirror

This noise induced behavior is called bubbling. In the case of chaos synchronized systems this bubbling leads to noise induced desynchronization. Depending on the overall structure of the phase space the trajectory may, after leaving the SM, approach another attractor or eventually return to the manifold.

Following [4], we will in this section discuss bubbling in a system of two delay coupled lasers.

13.1 Bubbling and On-off Intermittency in the Laser System

Consider two delay coupled lasers with delayed self-feedback, i.e., with the coupling scheme discussed in detail in Chap. 12. In particular we consider the case IVd from Table 12.1, where all delays are equal. For simplicity, we set all coupling phases to zero (compare Sect. 12.2.2) and the coupling coefficients equal to each other. Thus the laser equations are given by

$$\frac{d}{dt}E_j = \frac{1}{2}(1 + i\alpha)n_j E_j + \frac{1}{2}\kappa E_j(t - \tau) + \frac{1}{2}\kappa E_l(t - \tau) + F_E(t),$$

$$T\frac{d}{dt}n_j = p - n_j - (1 + n_j)|E_j|^2, \qquad (l = 3 - j).$$

This setup can experimentally be realized by coupling the lasers through a semitransparent mirror as depicted in Fig. 13.2. The coupling phases depend on the subwavelength tuning of the distances. Note that we consider the case where the mirror is positioned symmetrically between the lasers resulting in equal self-feedback and coupling delays. For a discussion of a system with an asymmetrically positioned mirror see [5, 6].

For this setup we calculated the maximum parallel LE and the maximum TLE (see Sect. 15.3 for a discussion of the numerical algorithm). The resulting exponents are depicted in Fig. 13.3 as a function of the coupling strengths κ.

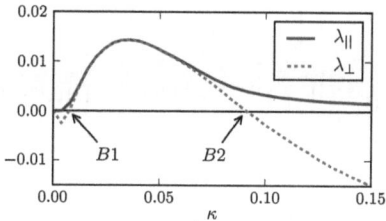

Fig. 13.3 Maximum transversal Lyapunov exponent (*dashed*) and maximum parallel Lyapunov exponent (*solid*) vs. feedback strengths κ. Parameters: $T = 200$, $p = 1.0$, $\tau = 1000$, $\alpha = 4.0$.

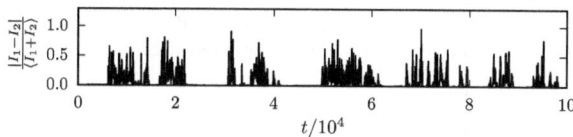

Fig. 13.4 On-off intermittency in the laser system. Plotted is the intensity difference $|I_1 - I_2|/\langle I_1 + I_2 \rangle$ normalized by the mean intensities (measuring the deviation from the synchronized state) vs time t. Periods of synchronization and desynchronization alternate. The noise was switched off in this simulation, which shows that the on-off intermittency is deterministic. Parameters: $\kappa = 0.085$ (close to 16.2) other parameters as in Fig. 13.3

The parallel exponent $\lambda_\parallel$ (*blue solid line*) is negative only for very small values of $\kappa \lesssim 0.004$ (this regime can only be seen when zooming in), i.e., the system is chaotic for κ above this threshold. The factor $1/2$ is included in the coupling strengths such that the synchronized dynamics is described by the LK equations (see Chap. 11) with a feedback strengths of κ.

The TLE $\lambda_\perp$ is depicted by the *red dashed line*. There are two blow-out bifurcations [2] at $\kappa \approx 0.008(16.1)$ and $\kappa \approx 0.09(16.2)$, where the TLE changes sign and the chaotic attractor loses its transversal stability. For κ values between 16.1 and 16.2 the chaotic attractor is transversely unstable. In this unstable regime we observe on-off intermittency close to the bifurcations, i.e., periods of synchronized and desynchronized motion alternate. This is depicted in Fig. 13.4. Note that this switching is a deterministic effect in contrast to bubbling. For values of κ further away from the blow-out bifurcations in the transversely unstable regime the periods of desynchronized motion become longer until the lasers show fully desynchronized behavior.

We now consider the range for κ above 16.2 in Fig. 13.3. As the TLE is negative in this range, the chaotic attractor is transversely stable and without any noise we observe perfect chaos synchronization. However, when taking the spontaneous emission noise into account, we observe desynchronization, which looks similar as on-off intermittency but is noise induced. Due to the negative TLE of the chaotic attractor the desynchronization must be due to bubbling and a

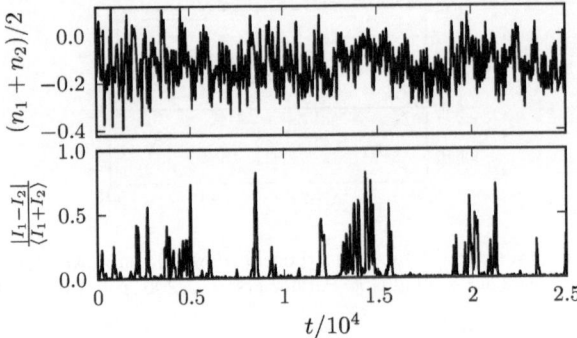

Fig. 13.5 Bubbling in the laser system in the coherence collapse regime. The top panel shows the symmetrized carrier density $(n_1 + n_2)/2$ vs. time and the bottom panel shows the intensity difference $|I_1 - I_2|/\langle I_1 + I_2 \rangle$ normalized by the mean intensities vs. time. The desynchronization is due to bubbling. Switching off the noise leads to perfect synchronization. Parameters: $T = 200$, $p = 1.0$, $\kappa = 0.12$, $\tau = 1000$, $\alpha = 4.0$

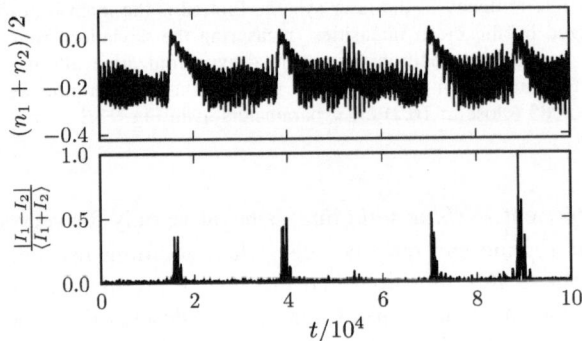

Fig. 13.6 Bubbling in the laser system in the low frequency fluctuation regime. The top panel shows the symmetrized carrier density $(n_1 + n_2)/2$ vs. time and the bottom panel shows the intensity difference $|I_1 - I_2|/\langle I_1 + I_2 \rangle$ normalized by the mean intensities vs. time. The desynchronization due to bubbling only occurs during power dropouts. Switching off the noise leads to perfect synchronization. Parameters as in Fig. 13.5 except $p = 0.1$

natural question to ask is which UPOs in the attractor are transversely unstable and thus responsible for the desynchronization.

A clue to answer this question can be found by investigating the particular desynchronization behavior in the two distinct chaotic regimes of the laser, namely the LFF-regime and the CC-regime. Figures 13.5 and 13.6 depict the dynamics in the CC and the LFF regime, respectively. The top panels depict the symmetrized carrier density $(n_1 + n_2)/2$ vs. time, which shows the chaotic dynamics of the lasers. In the lower panels the intensity difference $|I_1 - I_2|/\langle I_1 + I_2 \rangle$ normalized by the mean intensities is plotted vs time. In both cases bubbling occurs mainly at large carrier densities. This feature is very prominent in Fig. 13.6, where the desynchronization events occur at the power dropouts. We can thus presume that the transversely unstable orbits are located in a region of larger carrier density in the SM.

13.2 Transverse Stability of the Cavity Modes

As we will see the transverse stability of cavity modes in the SM plays a crucial role for the stability of the chaotic synchronization. We thus need to analyze this transverse stability.

As discussed in Chap. 12 the transversal stability is governed by the equation (12.10), which becomes in our case of equal coupling strengths

$$\frac{d}{dt}\delta A = Df(S)\delta A.$$

Thus for our particular setup of all coupling strengths being equal the delay term drops out and we only have to solve a linear time-dependent (since S depends on t) ordinary differential equation (ODE). When the dynamics $S(t)$ in the SM is given by a cavity mode

$$E(t) = E_* e^{i\omega t}, \quad n(t) = n_*$$

the matrix $Df[S(t)]$ becomes time-periodic. By transforming the laser coordinates into a co-rotating frame $\tilde{E} = E e^{-i\omega t}$ the cavity mode is transformed into a family of FPs

$$\tilde{E} = A e^{i\psi}, \quad n(t) = n_*$$

with ψ being the family parameter. All these FPs have the same stability as the PO in the initial coordinates and it is thus sufficient to analyze the stability of one of these FPs. Taking into account how the time derivative transforms into the co-rotating frame we obtain for our laser system the following linear equation, which governs the ECM's transverse stability

$$\begin{pmatrix} \delta x \\ \delta y \\ \delta n \end{pmatrix} = \begin{bmatrix} \frac{1}{2}n & -\frac{1}{2}\alpha n + \omega & \frac{1}{2}x - \frac{1}{2}\alpha y \\ \frac{1}{2}\alpha n - \omega & \frac{1}{2}n & \frac{1}{2}\alpha x + \frac{1}{2}y \\ -\frac{1}{T}(1+n)2x & -\frac{1}{T}(1+n)2y & -\frac{1}{T}(1+x^2+y^2) \end{bmatrix} \begin{pmatrix} \delta x \\ \delta y \\ \delta n \end{pmatrix},$$

where $\tilde{E} = x + iy$. For a given ECM with amplitude A, frequency ω and carrier density n we can set without loss of generality $x = A$ and $y = 0$ and calculate the eigenvalues of the matrix, which determine the transverse stability.

13.3 Relation Between Cavity Modes and Bubbling

By projecting the dynamics in the infinite dimensional synchronization manifold onto a two-dimensional plane spanned by the frequency $\omega = [\phi_s(t) - \phi_s(t-\tau)]/\tau$

Fig. 13.7 Projection of the
synchronized dynamics onto
a two dimensional plane.
Transversely stable and
unstable ECMs are shown as
blue circles and *red squares*,
respectively. Parameters:
$T = 200$, $p = 0.1$, $\kappa = 0.1$,
$\tau = 100$, $\alpha = 4.0$

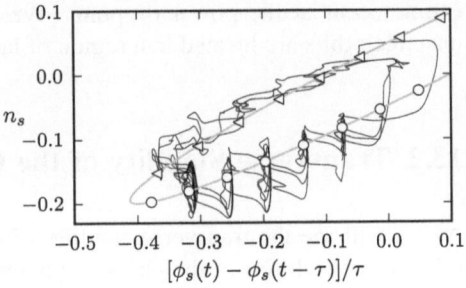

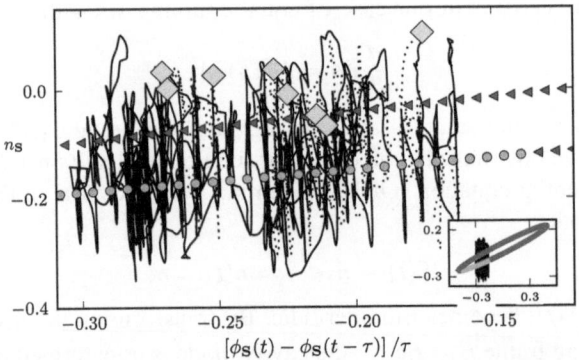

Fig. 13.8 Projection of the dynamics onto a two dimensional plane. Transversely stable and
unstable ECMs are shown as (*blue*) circles and (*red*) triangles, respectively. Diamonds (*yellow*)
mark the onset of desynchronization. *Solid* and *dashed* parts of the trajectory cor- respond to
synchronized and desynchronized periods, respectively. The inset in a shows the ECM ellipse
and bubbling dynamics in a larger range. Parameters: $T = 200$, $p = 1.0$, $\kappa = 0.12$, $\tau = 1000$,
$\alpha = 4.0$

and the symmetrized carrier density $n_s = (n_1 + n_2)/2$, where index s denotes that the
dynamics is in the synchronization manifold, we obtain Fig. 13.7. The transversely
stable and unstable cavity modes are shown as *blue circles* and *red squares*,
respectively. The modes involved during the power buildup process are all transversely stable and no bubbling occurs during this process. On the other hand during
the power dropouts, the anti-modes [7], which are all transversely unstable play a
crucial role and noise can induce bubbling. For this figure we used very small delay
time, which results in a small number of modes and gives a clearer picture of the
dynamics.

Figure 13.8 shows the same type of dynamics in the CC regime. As discussed in
Sect. 11.4, the dynamics in this regime is characterized by switching between
modes and crisis through collisions with anti-modes. This competition between
chaotic itinerancy and antimodes leads to bubbling during global antimode
dynamics. This can be seen by the location of the yellow diamonds, which mark
the onset of desynchronization. The desynchronization occurs when the chaotic

Fig. 13.9 Projection of the synchronized dynamics with an active relay onto a two dimensional plane. Transversely stable and unstable ECMs are shown as circles and triangles, respectively. Parameters: $T = 200$, $p_{relay} = 4$, $p_{outer} = 1.0$, $\kappa = 0.12$, $\tau = 1000$, $\alpha = 4.0$

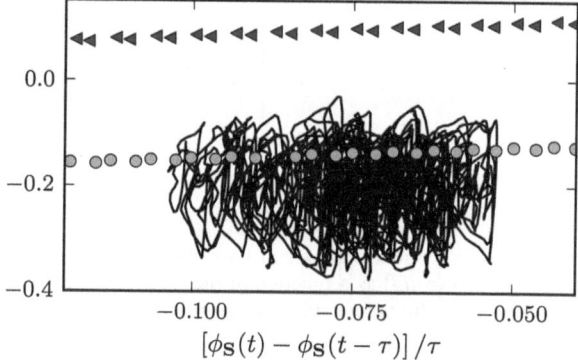

$$[\phi_{\mathrm{S}}(t) - \phi_{\mathrm{S}}(t - \tau)] / \tau$$

trajectory is close to the transversely unstable ECMs. Note that modes and antimodes are not necessarily transversely stable and unstable, respectively. The modes along the lower right side in Fig. 13.8, for instance, are transversely unstable.

With decreasing coupling strength κ, more modes become transversely unstable until the whole chaotic attractor loses its transversal stability. This leads to the blowout bifurcation 16.2 in Fig. 13.3. With increasing feedback strength the bubbling occurs less frequently and the average synchronization interval increases; however, we did not find a transition to a bubbling-free state in a physically reasonable range of K. This shows that the transverse stability of the cavity modes play a crucial role and determine the transverse stability of the chaotic orbit.

A natural question that arises is, whether there is a way to fully suppress the bubbling and lead to stable synchronization for the coupled lasers. As we have already mentioned above, we did not find any parameter ranges of the lasers, which lead to all ECMs involved in the dynamics being transversely stable, and thus we always observed bubbling.

However, when using another laser as an active relay between the outer lasers [4], we could suppress bubbling, when the relay laser was pumped stronger than the outer lasers. In the synchronized state the coupled system behaves like a single laser coupled bidirectionally to the relay laser, which corresponds to the case IIc in Table 12.1. For this system the chaotic dynamics [8] and the mode structure have been studied before [9, 10]. Here, the modes are called compound laser modes and have a more complex structure. Figure 13.8 shows the relevant part of the mode spectrum around which the system evolves. All modes in the proximity are transversely stable (*blue circles*) and bubbling is suppressed in this case. For this setup we have calculated the parallel and transversal Lyapunov exponent, too (see Fig. 13.10). Similarly, as in Fig. 13.3, we have two blow-out bifurcations 16.1 and 16.2. However, the transversal Lyapunov exponent quickly decreases to largely negative values with increasing κ, whereas the parallel exponent does not decrease as in Fig. 13.3, but instead increases. From this we can conclude, that the system is strongly chaotic and transversely more stable than the passive relay setup, which agrees with our results in Fig. 13.9.

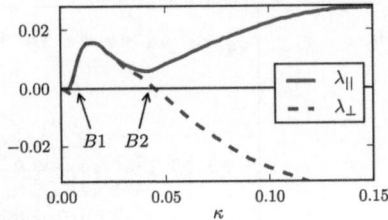

Fig. 13.10 Maximum transversal Lyapunov exponent (*dashed*) and maximum parallel Lyapunov exponent (*solid*) as a function of the feedback strength κ for an active. 16.1 and 16.2 mark two blow-out bifurcations. Parameters as in Fig. 13.9

13.4 Bubbling Statistics

Bubbling events can statistically be characterized through the maximum burst amplitude Δ and the average time between bursts T_b. It has been shown analytically [11, 12] by studying maps that there are two different transitions into the bubbling regime.

In the case of a *soft transition* the maximum burst height scales as $\Delta \propto \mu^{1/2}$, where μ is a normal parameter measuring the distance to the bubbling bifurcation and is zero at the bifurcation. For the *hard transition* the bursts set in with a finite maximum burst height $\Delta \propto O(1)$.

These scaling laws are only valid for scaling with respect to a *normal parameter* [13]. This is a parameter which does not change the dynamics within the SM but only the transverse stability of solutions in the SM. For the coupled laser system, there exist no such normal parameter, because every laser or coupling parameter also influences also the synchronized dynamics. However, we can conclude from numerical simulations that the transition should be a hard transition, for any of the varied parameters, because we do not observe a change in the burst height with changing parameters.

In the on-off intermittency regime, the statistical properties of the synchronization and desynchronization periods have also been studied analytically and numerically [14, 15]. Here, it was shown that the length L of synchronized periods has a power law distribution

$$P(L) \propto L^m.$$

The theory for maps predicts $m = -3/2$ [14]. We have calculated the distribution of synchronization lengths numerically for our laser model, the result is depicted in Fig. 13.11. We find a scaling with $m \approx -2.2$, which is not very good agreement with the theory. However, such deviations from the predictions from simple maps have previously been observed in this context [15, 16].

Fig. 13.11 Distribution of
synchronization lengths L in
the on-off intermittency
regime in log-log scale for the
delay coupled lasers. The
dashed straight line has a
slope of $m = -2.2$.
Parameters: $\kappa = 0.08$, $\alpha = 4$,
$T = 200$, $p = 1$, $\tau = 1000$

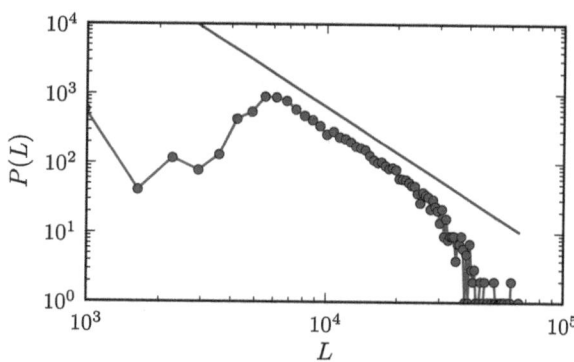

13.5 Basin of Attraction

The coupled laser system has an infinite dimensional phase space since it is a delay
system. It is therefore difficult to visualize the synchronization manifold's basin of
attraction. However it is possible to visualize two dimensional cuts of the basin
[17].

To do this we make a two dimensional cut through the space of history functions by choosing an ECM like history for each laser

$$E_1(t) = E_2(t) = A_0 e^{i\omega_0 t}$$
$$n_1(t) = n_1, \quad n_2(t) = n_2$$

and then vary the constants n_1 and n_2. For appropriately chosen parameters ω_0, A_0
and $n_1 = n_2$ this initial condition lies in the chaotic attractor in the synchronization
manifold.

If we now introduce small deviations between n_1 and n_2 the history lies close to
the synchronization manifold and we can track if such an initial condition results
in a bubbling event or if it decays quickly to the synchronization manifold again.

In the numerical simulations we simulated the lasers with the above initial
condition and calculated the time t_b it took for the intensities to deviate by 5 units

$$|I_1(t_b) - I_2(t_b)| = 5$$

If within a time of $t = 3200$ no bubbling occurred we stopped the simulation and
considered the initial condition to lie in the basin of attraction.

This method is shown in Fig. 13.12. Figure 13.13 depicts the result of the
simulation. The darker the color, the longer it takes for a bubbling event to occur.
Points in the dark blue region belong to the basin of attraction.

Note that the particular cut through the history function space as well as the
chosen threshold is somewhat arbitrary. It is only important that the two parameter
history function set intersects the basin of attraction and that the threshold has a
reasonable value. The basin seems to have a fractal structure, but more analysis
would be needed to confirm this conjecture. Such a fractal structure is typical for

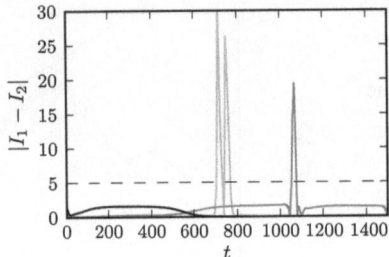

Fig. 13.12 Time series of $|I_1 - I_2|$ for three different histories. The *bright gray trajectories* display bubbling and go over the threshold of 5 (*dashed line*). The *dark gray trajectory* quickly decays and does not go over threshold line. The initial condition leading to this trajectory belongs to the basin of attraction. The colors correspond to the color code of Fig. 13.13

Fig. 13.13 Cut through the basin of attraction. The *color scale indicates* the time it takes for a bubbling event to occur (*dark* corresponds to long times). The dark blue region shows the basin of attraction

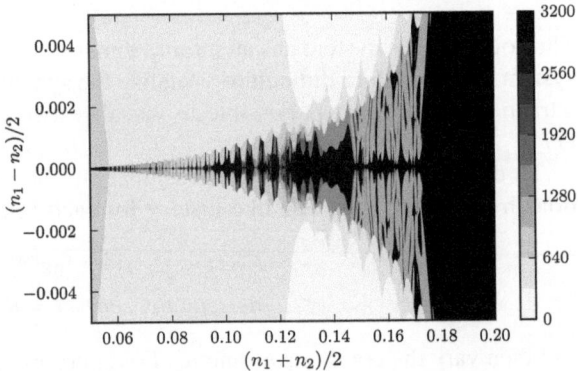

bubbling and on-off intermittency [2, 18] and the basin is called to be riddled in this case.

13.6 Conclusion

In this section we have discussed the occurrence of bubbling in a setup of two bidirectionally coupled lasers with self-feedback, corresponding to two lasers setup face to face with a semi-transparent mirror in between. We have seen that bubbling is always present in this setup and that it is caused by transversely unstable ECMs in the SM of the lasers. The particular location of the transversely unstable ECMs leads to an interesting interplay between bubbling and power dropouts in the LFF regime: during the power buildup process the modes involved in the chaotic itinerancy are transversely stable and lasers remain synchronized during this period; during the power dropout the trajectory collides with a transversely unstable antimode in a crisis and the lasers desynchronize due to bubbling.

We have seen that an active relay in form of another laser between the two outer lasers can suppress the bubbling. Furthermore, we have done some statistical analysis of the bubbling dynamics and investigated the basin of attraction.

References

1. P. Ashwin, J. Buescu, I. Stewart, Bubbling of attractors and synchronisation of chaotic oscillators. Phys. Lett. A **193**, 126 (1994)
2. E. Ott, J.C. Sommerer, Blowout bifurcations: the occurrence of riddled basins and on-off intermittency. Phys. Lett. A **188**, 39 (1994)
3. P. Cvitanović, R. Artuso, R. Mainieri, G. Tanner, G. Vattay, *Chaos: Classical and Quantum,* (Niels Bohr Institute, Copenhagen, 2008), http://ChaosBook.org
4. V. Flunkert, O. D'Huys, J. Danckaert, I. Fischer, E. Schöll, Bubbling in delay-coupled lasers. Phys. Rev. E **79**, 065201 R (2009)
5. K. Hicke, *Stability of synchronized states in delay coupled lasers*, Master's thesis, TU, Berlin, (2009)
6. K. Hicke, O. D'Huys, V. Flunkert, E. Schöll, J. Danckaert, I. Fischer, Mismatch and synchronization: Influence of asymmetries in systems of two delay-coupled lasers. Phys. Rev. E **83**, 056211 (2011)
7. T. Sano, Antimode dynamics and chaotic itinerancy in the coherence collapse of semiconductor lasers with optical feedback. Phys. Rev. A **50**, 2719 (1994)
8. J. Mulet, C.R. Mirasso, T. Heil, I. Fischer, Synchronization scenario of two distant mutually coupled semiconductor lasers. J. Opt. B **6**, 97 (2004)
9. H. Erzgräber, D. Lenstra, B. Krauskopf, E. Wille, M. Peil, I. Fischer, W. Elsäßer, Mutually delay-coupled semiconductor lasers: Mode bifurcation scenarios. Opt. Commun. **255**, 286 (2005)
10. H. Erzgräber, B. Krauskopf, D. Lenstra, Compound laser modes of mutually delay-coupled lasers. SIAM J. Appl. Dyn. Syst. **5**, 30 (2006)
11. S.C. Venkataramani, B.R. Hunt, E. Ott, Bubbling transition. Phys. Rev. E **54**, 1346 (1996)
12. S.C. Venkataramani, B.R. Hunt, E. Ott, D.J. Gauthier, J.C. Bienfang, Transitions to bubbling of chaotic systems. Phys. Rev. Lett. **77**, 5361 (1996)
13. J.R. Terry, K.S. Thornburg, D.J. DeShazer, G.D. Van Wiggeren, S. Zhu, P. Ashwin, R. Roy, Synchronization of chaos in an array of three lasers. Phys. Rev. E **59**, 4036 (1999)
14. J.F. Heagy, N. Platt, S. Hammel, Characterization of on-off intermittency. Phys. Rev. E **49**, 1140 (1994)
15. M. Sauer, F. Kaiser, On-off intermittency and bubbling in the synchronization break-down of coupled lasers. Phys. Lett. A **243**, 38 (1998)
16. J. Redondo, E. Roldán, G. de Valcarcel, On-off intermittency in a Zeeman laser model. Phys. Lett. A **210**, 301 (1996)
17. S.R. Taylor, S.A. Campbell, Approximating chaotic saddles for delay differential equations. Phys. Rev. E **75**, 46215 (2007)
18. E. Ott, *Chaos in Dynamical Systems.* (Cambridge University Press, Cambridge, 1993)

Chapter 14
Summary and Conclusions

In this thesis I investigated complex systems under the influence of time delay. The first part dealt with time-delayed feedback control and more specifically the stabilization of odd-number orbits by time-delayed feedback control. I discussed in detail the counter example which refutes the alleged odd-number theorem. Furthermore, I considered new feedback schemes, which are motivated by the experimentally most relevant situation: having access to one measurement variable and being able to apply a control signal to another input variable of the system. In this context I also considered symmetric feedback matrices, which could previously not stabilize odd-number orbits. I showed that by introducing an additional latency in the control loop, this difficulty can be overcome and stabilization is possible. As an application of these new feedback schemes I showed that in a laser model, which exhibits a subcritical Hopf bifurcation, the subcritical orbit can be stabilized using optoelectronic feedback of Pyragas type.

Besides individual systems with time-delayed feedback, I considered diffusively coupled normal form oscillators and the stabilization of odd-number orbits, corresponding to in-phase or anti-phase solutions, by time-delayed coupling—a generalization of Pyragas feedback.

The second part of this work was devoted to synchronization phenomena in delay-coupled systems. I analyzed networks with delayed-connections using the master stability function approach and showed that the master stability function has a simple structure in the limit of large delays—large in comparison with the internal time scale of the nodes. This structure allowed me to draw very general conclusions about the synchronizability of network structures. For the proof I extended a scaling theory for large delay which was developed in the context for flows to the case of delayed maps.

From these general considerations I continued with the analysis of simple coupling topologies: network motifs of two delay coupled elements. Here I derived necessary conditions on the coupling parameters which guarantee the existence of an invariant synchronized solution. For laser systems which have an internal S^1-symmetry these coupling conditions lead to non-trivial conditions on the coupling phases.

V. Flunkert, *Delay-Coupled Complex Systems*, Springer Theses,
DOI: 10.1007/978-3-642-20250-6_14, © Springer-Verlag Berlin Heidelberg 2011

I then focused on one of the previously discussed network motifs: two lasers with delayed self-feedback and coupling, where all feedback parameters are equal. For this model one observes bubbling and on-off intermittency, which I could explain by stability features of the system's unstable laser modes.

For the efficient simulation of the delay differential equations in this thesis, I developed a software tool, which is discussed together with the numerical and analytical methods in the appendix.

I will now discuss some open questions and directions for future works. Until now, odd-number orbits have been studied only close to bifurcations, where one can analyze the system using normal forms and center manifold theory. In many applications, however, the systems are operating far away from bifurcations. It is still unclear whether stabilization can be achieved in these situations and how appropriate feedback matrices can be constructed. Furthermore, at the time of writing no experiment concerning the stabilization of an odd-number orbit has yet been published, although different groups are working on such experiments. This is an important short-term objective since it will confirm the refutation of the odd-number theorem experimentally. Perhaps the experimentally relevant feedback schemes discussed in this work could prove to be helpful.

Concerning delay coupled systems and networks with coupling delays, an important question is what effects multiple delays and distributed delays have in these systems. The delays occurring for example in neural networks depend on the distance of the cells and other parameters and are thus distributed or even stochastic. In such circumstances one cannot expect complete synchronization, however, partial synchronization effects are known to play a crucial role in the brain. The analysis of chimera states is a first step in the understanding of such systems at the brink of synchronization. But simple analytic tools such as the master stability function approach fail in these situations and new methods have to be developed.

Part III
Appendix

Chapter 15
Delay Differential Equations

Delay differential equations occur in many areas of science. Mathematically, delay terms render differential equations infinite dimensional. This enables even simple equations with delay terms to show complex dynamics. As such a simple example consider the Mackey-Glass equation

$$\frac{d}{dt}x(t) = \beta \frac{x(t-\tau)}{1+x(t-\tau)^p} - \gamma x(t),$$

which is one of the earliest and best studied delay equation. It describes the white blood cell concentration x(t) in the blood [1] and is a simple rate equation: it has a linear decay term $-\gamma x(t)$ describing the depletion of white blood cells and a nonlinear production term which reacts to the existing concentration with a finite reaction time τ. Already this basic model shows very interesting dynamics including oscillations, chaos and multistability.

15.1 Numerical Simulation of Differential Equations

For the numerical simulation of DDEs special tailored algorithms have been developed. We will discuss the Bogacki-Shampine method for delay equations, which is one of the most widely used algorithms.

For the convenient simulation of DDEs the algorithm has been implemented in a simulation package [2] (see Chap. 16).

15.1.1 Bogacki-Shampine Method

Runge-Kutta (RK) methods are among the most important methods for solving ODEs. One widely used RK method is the Bogacki-Shampine method. For an initial value problem

V. Flunkert, *Delay-Coupled Complex Systems*, Springer Theses,
DOI: 10.1007/978-3-642-20250-6_15, © Springer-Verlag Berlin Heidelberg 2011

$$\frac{d}{dt}X(t) = f(t, X(t)), \quad X(t_0) = X_0 \tag{15.1}$$

the Bogacki-Shampine method is calculated according to

$$X_{n+1} = X_n + h_n \frac{1}{9}(2k_1 + 3k_2 + 4k_3),$$

$$t_{n+1} = t_n + h_n$$

with the step size h_n and

$$k_1 = f(t_n, X_n),$$

$$k_2 = f(t_n + \frac{1}{2}h_n, X_n + \frac{1}{2}h_n k_1),$$

$$k_3 = f(t_n + \frac{3}{4}h_n, X_n + \frac{3}{4}h_n k_2),$$

starting with X_0 at t_0.

The solution X(t) is approximated by the values X_n at the sampling points t_n. The distance between the sampling points is given by the step size h_n. The Bogacki-Shampine algorithm is a third order method, meaning that the error made in each step is of the order h^3. Another second order approximation comes at very little computational expense with the Bogacki-Shampine algorithm

$$\tilde{X}_{n+1} = X_n + h_n \frac{1}{24}(7k_1 + 6k_2 + 8k_3 + 3k_4),$$

$$k_4 = f(t + h_n, X_{n+1}).$$

This allows an error estimation

$$e_{n+1} = X_{n+1} - \tilde{X}_{n+1}$$

$$= X_n + h_n \frac{1}{9}(2k_1 + 3k_2 + 4k_3) - \left[X_n + h_n \frac{1}{24}(7k_1 + 6k_2 + 8k_3 + 3k_4)\right]$$

$$= h_n \frac{1}{72}(-5k_1 + 6k_2 + 8k_3 - 9k_4).$$

If the error e_{n+1} becomes too large the step can be repeated with a smaller step size until the error is small enough.

A possible criterion for accepting a step is

$$|e| \leq \max(\text{RelTol} \cdot |X|, \text{AbsTol})$$

with the relative tolerance $\text{RelTol} = 10^{-3}$ and the absolute tolerance $\text{AbsTol} = 10^{-6}$. This method with an adaptive step size correction is implemented in Matlab as the ode23 function.

15.1.2 Runge-Kutta Method for Delay Differential Equations

For RK methods it is essential to calculate f at intermittent points between the actual sampling points, e.g., $f(t_n + \frac{1}{2}h_n, X_n + \frac{1}{2}h_n k_1)$. This is the main problem in generalizing RK algorithms to DDEs [3].

For a DDE with a history function $\phi(t)$

$$\frac{d}{dt}X(t) = f(t, X(t), X(t - \tau)), \quad X(t) = \phi(t) \text{ for } t \in [-\tau, 0] \tag{15.2}$$

it is easy to apply the Bogacki-Shampine algorithm for the interval $t \in [0, \tau]$ since on this interval (15.2) is an ODE when $X(t - \tau)$ is replaced by the known history function $\phi(t)$. One obtains the approximation of $X(t)$ on the sampling points in the interval $[0, \tau]$. On the next interval $t \in [\tau, 2\tau]$ the history $X(t - \tau)$ is only known on these sampling points. In order to calculating f on intermittent points it is now necessary to interpolate $X(t)$ in between the sampling points, i.e., we need the values of $X(t_n - \tau + \frac{1}{2}h_n)$. This becomes even more important when using adaptive step size methods.

There are now different possibilities to interpolate $X(t)$ on intermittent points. The algorithm's order of accuracy is determined by the order of the RK method as well as the order of the interpolation—ideally, the orders are the same.

A third order interpolation of the history between two sampling times t_n and t_{n+1} can be done by using the values $X(t_n)$ and $X(t_{n+1})$ and the derivatives $X'(t_n)$ and $X'(t_{n+1})$ which have to be saved during the simulation. Making a (vector valued) polynomial ansatz

$$X_{\text{interp}}(t) = a_0 + a_1 t + a_2 t^2 + a_3 t^3 \tag{15.3}$$

the (vector valued) coefficients can be explicitly calculated from

$$X(t_n) = a_0 + a_1 t_n + a_2 t_n^2 + a_3 t_n^3$$
$$X(t_{n+1}) = a_0 + a_1 t_{n+1} + a_2 t_{n+1}^2 + a_3 t_{n+1}^3$$
$$X'(t_n) = a_1 + 2a_2 t_n + 3a_3 t_n^2$$
$$X'(t_{n+1}) = a_1 + 2a_2 t_{n+1} + 3a_3 t_{n+1}^2.$$

This then yields the interpolation of X between t_n and t_{n+1}

$$X_{\text{interp}}(t) = \frac{1}{(t_n - t_{n+1})^3} \Big[(t - t_n)(t - t_{n+1})^2 (t_n - t_{n+1})X'(t_n)$$
$$+ (t - t_n)^2 (t - t_{n+1})(t_n - t_{n+1})X'(t_{n+1})$$
$$- (t - t_{n+1})^2 (2t - 3t_n + t_{n+1})X(t_n)$$
$$+ (t - t_n)^2 (2t + t_n - 3t_{n+1})X(t_{n+1}) \Big]. \tag{15.4}$$

15.1.2.1 Constant Step Size

For constant step size $t_{n+1} - t_n = h_n = h$ and the delay being an integer multiple of the step size $\tau = vh$, one can simplify the calculation further. The Bogacki-Shampine methods needs the values of X_{interp} at $t_n - \tau + \frac{1}{2}h$ and $t_n - \tau + \frac{3}{4}h$. Using these values in (15.4) yields

$$X_{\text{interp}}\left(t_n - \tau + \frac{1}{2}h\right) = \frac{h}{8}[X'(t_{n-v}) - X'(t_{n+1-v})] + \frac{1}{2}[X(t_{n-v}) + X(t_{n+1-v})],$$

$$X_{\text{interp}}\left(t_n - \tau + \frac{3}{4}h\right) = \frac{h}{64}[3X'(t_{n-v}) - 9X'(t_{n+1-v})] + \frac{1}{32}[5X(t_{n-v}) + 27X(t_{n+1-v})],$$

where we used $t_n - \tau = t_{n-v}$.

In this case the DDE

$$\frac{d}{dt}X(t) = f(t, X(t), X(t - \tau))$$

can be simulated using the Bogacki-Shampine method by

$$X_{n+1} = X_n + h\frac{1}{9}(2k_1 + 3k_2 + 4k_3),$$

$$t_{n+1} = t_n + h$$

with

$$k_1 = f(t_n, X_n, X_{n-v}),$$

$$k_2 = f(t_n + \frac{1}{2}h, X_n + \frac{1}{2}hk_1, X_{\text{interp}}(t_{n-v} + \frac{1}{2}h)),$$

$$k_3 = f(t_n + \frac{3}{4}h, X_n + \frac{3}{4}hk_2, X_{\text{interp}}(t_{n-v} + \frac{3}{4}h)).$$

15.1.2.2 Variable Step Size

For variable step size the interpolation (15.4) can not be simplified further. In the RK steps for each evaluation of X_{interp} at some value s one has to first find the sampling points in the history t_m and t_{m+1}, such that

$$t_m \leq s \leq t_{m+1}$$

and then use the interpolation using these sampling points. Since the sampling points are not equidistant the time points at $t - \tau + \frac{1}{2}h$ and $t - \tau + \frac{3}{4}h$ do not necessarily lie between the same sampling points.

15.1.2.3 Noise

For the Euler method it is known how noise can be handled in the numerical simulation. The noise amplitude scales with the square root of the step size $\sqrt{\Delta t}$. For higher order method the adequate handling of noise terms is a very delicate business not to mention higher order methods for DDEs with noise. For this reason a pragmatic approach is often used in practice. The deterministic parts of the equations are handled using a high order method as described above. Finally, the noise is added to each step implemented via the Euler method. Here it is important that the noise realization enters the dynamics in the following step and is not just an "observation noise".

15.2 Floquet Exponents

Floquet exponents describe the stability of a PO in a dynamical system. Consider an n-dimensional autonomous ODE

$$\frac{d}{dt}X(t) = f(X(t)), \qquad (X \in \mathbb{R}^n) \tag{15.5}$$

with a PO $X_*(t)$ with period T, i.e., $X_*(t)$ is a solution of (15.5) with

$$X_*(t + T) = X_*(t).$$

To determine the stability of this orbit, we consider a perturbation $\delta x(t)$ to the PO $X(t) = X_*(t) + \delta x(t)$ and linearize (15.5) in δx around the PO. This yields a variational equation for the perturbation

$$\frac{d}{dt}\delta x(t) = A(t)\delta x(t), \tag{15.6}$$

where $A(t) := Df(X_*(t))$ is the Jacobian matrix along the PO. Since (15.6) is a linear equation, albeit time-dependent, the superposition principle applies. To solve (15.6) for any initial condition, we can use the fundamental matrix $\Phi(t, t_0)$, which solves

$$\frac{d}{dt}\Phi(t, t_0) = A(t)\Phi(t, t_0), \quad \text{and} \quad \Phi(t_0, t_0) = I_n,$$

where I_n is the $n \times n$ identity matrix. Formally, the fundamental matrix is given by the time-ordered integral

$$\Phi(t, t_0) = \mathbf{T}e^{\int_{t_0}^{t} ds A(s)}.$$

An initial perturbation $\delta x(t_0)$ then evolves according to

$$\delta x(t) = \Phi(t, t_0)\delta x(t_0).$$

The PO X_* is stable if any such perturbation does not grow.

Equation (15.6) is T-periodic and it can be shown that the fundamental matrix is also T-periodic

$$\Phi(t + T, t_0 + T) = \Phi(t, t_0).$$

It is therefore sufficient to consider the fundamental matrix after one period

$$M(t_0) := \Phi(t_0 + T, t_0).$$

This matrix is called the PO's monodromy matrix. The eigenvalues $\mu_j (j = 1, \ldots, n)$ of the monodromy matrix are independent of the starting point on the PO and topologically invariant, i.e., independent of the coordinate system. With the eigenvectors[1] $\mathbf{e}_j$ the eigenvalue equation may be written as

$$M(t_0)\mathbf{e_j} = \mu_j \mathbf{e_j} = e^{\Lambda_j T}\mathbf{e_j},$$

where the imaginary part of the complex number $\Lambda_j T$ is only defined modulo 2π and can, for instance, be chosen to lie in $[0, 2\pi)$.

Let us now use time-dependent coefficients $c_j(t)$ to expand a perturbation $\delta x(t)$ using the vectors $\mathbf{e}_j$

$$\delta x(t) = \sum_j c_j(t)\mathbf{e_j}.$$

What happens to this perturbations after one period? Using the fundamental matrix we have

$$\delta x(t_0 + T) = \Phi(t_0 + T, t_0)\delta x(t_0) = M(t_0)\sum_j c_j(t)\mathbf{e_j} = \sum_j c_j(t)e^{\Lambda_j T}\mathbf{e_j}.$$

On the other hand

$$\delta x(t_0 + T) = \sum_j c_j(t_0 + T)\mathbf{e_j},$$

i.e.,

$$c_j(t_0 + T) = c_j(t_0)e^{\Lambda_j T}$$

for all t_0. The coefficients can hence be split into two parts: a T-periodic part $v_j(t)$ and an exponential part $e^{\Lambda_j t}$:

---

[1] Note that we only use a bold typeface for these eigenvectors and do not consistently write all vectors in bold.

$$c_j(t) = v_j(t)e^{\Lambda_j t}.$$

This gives the Floquet theorem for differential equations: Any solution $\delta x(t)$ of (15.6) may be expanded as

$$\delta x(t) = \sum_j v_j(t)e^{\Lambda_j t}\mathbf{e_j},$$

where $v_j(t)$ are T-periodic. The complex numbers Λ_j and $\mu_j = e^{\Lambda_j T}$ are called the Floquet exponents and Floquet multipliers of the PO, respectively. As discussed above, the imaginary part of each Floquet exponent is only defined modulo $2\pi/T$.

For autonomous ODEs one of the Floquet multipliers is unity and the corresponding Floquet exponent is zero. This corresponds to an initial perturbation $\delta x(t_0)$ tangent to the periodic orbit. This perturbation is left unchanged during the time evolution. This so-called Goldstone mode is a result of the system's symmetry under time-shifts.

A periodic orbit is asymptotically stable if all Floquet exponents except the Goldstone mode have real part smaller than zero, or equivalently, all Floquet multipliers have magnitude smaller than one except the Goldstone mode. This implies that a perturbation to the PO decays, except for its component tangent to the orbit, which remains as a constant phase shift.

15.3 Lyapunov Exponents

Lyapunov Exponents (LEs) determine whether a dynamical system is chaotic and quantify the strength of the chaos. Consider a dynamical system

$$\frac{d}{dt}x(t) = f(x). \tag{15.7}$$

With the flow $h^t(\cdot)$ of the dynamical system the evolution of two trajectories starting with an initial (infinitesimally small) separation $\delta x(0) = \delta x_0$ are given by

$$x(t) = h^t(x_0) \quad \text{and} \quad x(t) + \delta x(t) = h^t(x_0 + \delta x_0).$$

If the system is chaotic, any two such trajectories separate exponentially fast

$$\delta x(t) \approx e^{\lambda t}\delta x_0.$$

The rate λ is called the LE and is positive in this case.

Note that a positive LE is not a sufficient condition for chaos, since a non-chaotic repeller also results in exponential divergence of neighboring trajectories. An exact definition of chaos is as follows [4]:

- The system must be sensitive to initial conditions (positive LE).
- The system must be topologically mixing.

Fig. 15.1 Growth of
variations along independent
directions

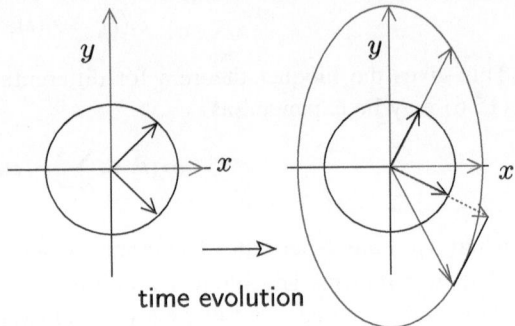

time evolution

- There must be an infinite number of periodic orbits which are dense in the attractor.

However, for all practical purposes it is sufficient to have a positive LE and to require that the dynamics is bounded.

The rate of divergence ($\lambda > 0$) or convergence ($\lambda < 0$) depends on the orientation of the initial separation δx_0 and the number of independent Lyapunov exponents is equal to the number of dimensions of the phase space. The set of all Lyapunov exponents is called the Lyapunov spectrum.

As long as the initial separation δx_0 has a component along the maximum growth direction the maximum Lyapunov exponent will dominate the growth. Thus for almost all initial separations the leading exponent is given by

$$\lambda = \lim_{t \to \infty} \ln \frac{\|\delta x(t)\|}{\|\delta x_0\|}.$$

To find not only the largest exponent but the leading N exponents a Gram-Schmidt procedure can be used. This is depicted in Fig. 15.1. The initial variations (black arrows) grow during time evolution (blue solid arrows). The largest Lyapunov exponent (y-direction) dominates this growth for all variations which have a component in that direction. A Gram-Schmidt procedure orthogonalizes the variations (blue dotted arrow) and gives a better estimate of the growth in the x-direction. After normalization (black arrows) a new iteration of the algorithm starts. With each iterations the variations align better along the independent growth directions.

A number of methods [5] have been developed to calculated Lyapunov exponents from numerically or experimentally obtained time traces. The main idea is to use a delay embedding in a space of large enough dimension. If the embedding dimensions is larger than the dimensionality of the dynamics and the delay coordinates are chosen properly, points which are close together in the embedding space are also close in the phase space of the dynamical system. The divergence of such neighboring points can then be averaged and interpolated by an exponential law, yielding the desired Lyapunov exponent.

These algorithms are the method of choice for experimental data. However, for numerical simulations better methods exist, since the underlying dynamical equations are known.

For infinitesimal small initial separations δx_0 the evolution of the separation $\delta x(t)$ from the trajectories $x(t)$ is governed by the variational equation

$$\frac{d}{dt}\delta x(t) = Df(x(t))\delta x(t), \tag{15.8}$$

where $D f(x(t))$ is the Jacobian of f evaluated at x(t). Thus, one can directly simulate the linear equation for the separation (15.8) together with the system (15.7) instead of simulating (15.7) for the two neighboring initial conditions. The advantage of this calculation in the tangent space is that one is always in the linear regime no matter how large the tangent vector δx becomes.

To find not only the leading exponent but the leading N exponents the following algorithm can be employed [6].

1. Choose $i = 1, 2, \ldots, N$ orthonormal vectors $\delta x_0^{(i)}$.
2. Simulate the system (15.7) and the variational (15.8) for each $\delta x_0^{(i)}$ over an appropriate period T.
3. Orthogonalize the vectors $\delta x^{(i)}(T)$ using a Gram-Schmidt procedure to obtain the vectors $\overline{\delta x^{(i)}(T)}$ (no normalization).
4. An estimate for the ith Lyapunov exponent is given by

$$\lambda^{(i)} = \ln \frac{\|\overline{\delta x^{(i)}(T)}\|}{\|\delta x_0^{(i)}\|} = \ln \|\overline{\delta x^{(i)}(T)}\|.$$

5. Normalize the $\overline{\delta x^{(i)}(T)}$ and start from 2. with these vectors.

Each repetition of steps 2–5 yields an estimate of the N leading Lyapunov exponents. To obtain accurate values and error bounds for the Lyapunov exponents we can calculate the mean and standard error of the mean of the obtained $\lambda^{(i)}$s.

The time interval T after which the estimate of the Lyapunov exponents are calculated and the vectors are orthonormalized should neither be too short nor too long. Too short T's will result in the accumulation of round-off errors. Too long T's will lead to the variations $\delta x^{(i)}$ becoming very large and thus a loss of numerical precision. Best results are obtained if T is of the order of the system time scale. For delay differential equations it is convenient to use $T = \tau$.

15.3.1 Lyapunov Exponents for Systems with Delay

The above algorithm has been generalized in [7] to the case of delay differential equations.

Consider a system governed by the delay-differential equation

$$\frac{d}{dt}x(t) = f(x, x_\tau).$$ (15.9)

The state vector of such a system with delay is a function $x(t)$ over a time interval $[t - \tau, t]$. The evolution of an infinitesimal separation δx is governed in the delay-differential case by the equation

$$\frac{d}{dt}\delta x = D_x f(x, x_\tau)\delta x + D_{x_\tau} f(x, x_\tau)\delta x_\tau,$$ (15.10)

where the $D_x f$ and $D_{x_\tau} f$ represent the Jacobian matrices of f with respect to x and x_τ. Here, the variations are functions over the time interval $[t - \tau, t]$.

The idea of Farmer [7] is to discretize the function $x(t)$ as well as the separations δx over the interval $[t - \tau, t]$ to obtain a finite dimensional approximation for the delay-differential equation. Discretizing the time as $t_k := k\Delta t$, where $\Delta t = \tau/(n - 1)$ is the sampling size, the functions $x(t)$ and $\delta x(t)$ are sampled on the interval $[t - \tau, t]$ by n samples.
We use the notations

$$x_k := x(t_k), \; J_k := D_x f(x(t_k), x(t_k - \tau)) \; J_k^{(\tau)} := D_{x_\tau} f(x(t_k), x(t_k - \tau))$$

for simplicity.

The DDE can be integrated on the sampling points using an Euler scheme

$$x_{k+1} = x_k + \Delta t f\big(x_k, x_{k-(n-1)}\big).$$

Similarly the variational equation can be integrated by

$$\delta x_{k+1} = \delta x_k + \Delta t\Big[J_k \delta x_k + J_k^{(\tau)} \delta x_{k-(n-1)}\Big].$$

Note that J_k and $J_k^{(\tau)}$ depend on the x_k and $x_{k-(n-1)}$, i.e., the equation for x has to be integrated along side with the variational equation. In practice a higher order method should be used to integrate the differential equations.

Using the vector

$$\delta\psi_k := (\delta x_k, \delta x_{k-1}, \ldots, \delta x_{k-(n-1)})$$

the integration scheme for the variational equation can also be written as

$$\delta\psi_{k+1} = \delta\psi_k + \Delta t
\begin{bmatrix}
J_k & & & & J_k^{(\tau)} \\
1 & 0 & & & \\
& \ddots & \ddots & & \\
& & & 1 & 0
\end{bmatrix}
\delta\psi_k.$$

Since the space of vectors $\delta\psi_k$ is just an $n \cdot \mathrm{Dim}(\delta x)$ dimensional real vector space, we can apply the ordinary algorithm to calculate the leading N Lyapunov exponents.

References

1. M.C. Mackey, L. Glass, Oscillation and chaos in physiological control systems. Science **197**, 287 (1977)
2. V. Flunkert, E. Schöll, pydelay—a python tool for solving delay differential equations. (2009) arXiv:0911.1633 [nlin.CD]
3. C. Baker, C. Paul, D. Willé, Issues in the numerical solution of evolutionary delay differential equations. Adv. Comput. Math. **3**, 171 (1995)
4. A. Katok, B. Hasselblatt, Introduction to the modern theory of dynamical systems. (Cambridge University Press, Cambridge,MA, 1996)
5. A. Wolf, J.B. Swift, H.L. Swinney, J.A. Vastano, Determining Lyapunov exponents from a time series. Physica D **16**, 285 (1985)
6. G. Benettin, L. Galgani, A. Giorgilli, J. Strelcyn, Lyapunov characteristic exponents for smooth dynamical systems; a method for computing them all. Meccanica **15**, 9 (1980)
7. J.D. Farmer, Chaotic attractors of an infinite-dimensional dynamical system. Physica D **4**, 366 (1982)

Chapter 16
Pydelay: A Simulation Package

16.1 Introduction

Pydelay is a program (licensed under the MIT license) which translates a system
of DDEs into simulation C-code and compiles and runs the code (using scipy
weave) [1]. This way it is easy to quickly implement a system of DDEs but you
still have the speed of C. The Homepage can be found here:

http://pydelay.sourceforge.net/

It is largely inspired by PyDSTool.

The algorithm used is based on the Bogacki-Shampine method [2] which is also
implemented in Matlab's dde23 [3].

I also want to mention PyDDE—a different python program for solving DDEs.

16.1.1 Installation and Requirements

Unix: You need python and python headers files (in debian/ubuntu these are in the
package python-dev), numpy and scipy and the gcc-compiler.

To plot the solutions and run the examples you also need matplotlib.

To install pydelay download the latest tar.gz from the website and install the
package in the usual way:

```
cd pydelay-$version
python setup.py install
```

When the package is installed, you can get some info about the functions and the
usage with:

```
pydoc pydelay
```

V. Flunkert, *Delay-Coupled Complex Systems*, Springer Theses,
DOI: 10.1007/978-3-642-20250-6_16, © Springer-Verlag Berlin Heidelberg 2011

For Arch linux there is a PKGBUILD.

Windows: The solver has not been tested on a `windows` machine. It could perhaps work under cygwin.

16.1.2 An Example

The following example shows the basic usage. It solves the Mackey-Glass equation [4]

$$\dot{x} = \beta \frac{x(t-\tau)}{1 + x(t-\tau)^p} - \gamma x$$

for parameters and initial conditions which lead to a periodic orbit.[1]

```python
# import pydelay and numpy and pylab
import numpy as np
import pylab as pl
from pydelay import dde23

# define the equations
eqns = {
    'x' : 'beta*x(t-tau) / (1.0 + pow(x(t-tau),p)) - gamma*x'
    }

#define the parameters
params = {
    'tau'  : 15,
    'p'    : 10,
    'beta' : 0.25,
    'gamma': 0.1
    }

# Initialise the solver
dde = dde23(eqns=eqns, params=params)

# set the simulation parameters
# (solve from t=0 to t=1000 and
# limit the maximum step size to 1.0)
dde.set_sim_params(tfinal=1000, dtmax=1.0)
```

---

[1] See http://www.scholarpedia.org/article/Mackey-Glass_equation for this example.

```
# set the history of to the constant
# function 0.5 (using a python lambda function)
histfunc = {
    'x': lambda t: 0.5
    }
dde.hist_from_funcs(histfunc, 51)

# run the simulator
dde.run()

# Make a plot of x(t) vs x(t-tau):
# Sample the solution twice with a stepsize of dt=0.1:
# once in the interval [515, 1000]
sol1 = dde.sample(515, 1000, 0.1)
x1 = sol1['x']

# and once between [500, 1000-15]
sol2 = dde.sample(500, 1000-15, 0.1)
x2 = sol2['x']

pl.plot(x1, x2)
pl.xlabel('$x(t)$')
pl.ylabel('$x(t - 15)$')
pl.show()
```

Figure 16.1 shows the resulting plot.

16.2 Usage

16.2.1 Defining the Equations, Delays and Parameters

Equations are defined using a python dictionary. The keys are the variable names and the entry is the right hand side of the differential equation. The string defining

Fig. 16.1 Periodic orbit in the Mackey-Glass model

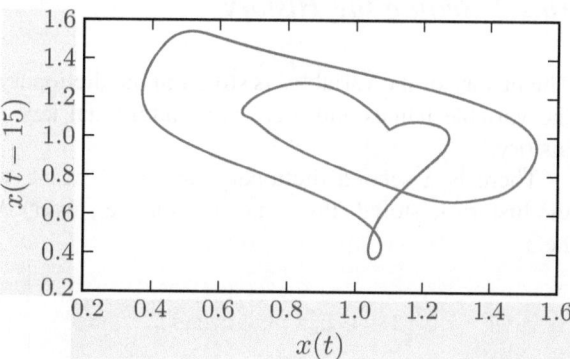

the equation has to be a valid C expression, i.e., use `pow(a,b)` instead of `a**b` etc.

Delays are written as `(t-delay)`, where `delay` can be some expression involving parameters and numbers but not (yet) involving the time `t` or the dynamic variables:

```
eqns = {
    'y1': '- y1 * y2(t-tau) + y2(t-1.0)',
    'y2': 'a * y1 * y2(t-2*tau) - y2',
    'y3': 'y2 - y2(t-(tau+1))'
}
```

Complex variables can be defined by adding `':c'` or `':C'` in the eqn-dictionary. The imaginary unit can be used through `'ii'` in the equations:

```
eqns = {
    'z:c': '(la + ii*w0 + g*pow(abs(z),2) )*z
            + b*(z(t-tau) - z(t))',
}
```

Parameters are defined in a separate dictionary where the keys are the parameter names, i.e.,:

```
params = {
    'a'  : 0.2,
    'tau': 1.0
}
```

16.2.2 Setting the History

The history of the variables is stored in the dictionary `dde23.hist`. The keys are the variable names and there is an additional key `'t'` for the time array of the history.

There is a second dictionary `dde23.Vhist` where the time derivatives of the history is stored (this is needed for the solver). When the solver is initialized, i.e.,:

```
dde = dde23(eqns, params)
```

the history of all variables (defined in eqns) is initialized to an array of length nn = 101 filled with zeros. The time array is evenly spaced in the interval [-maxdelay, 0].

It is possible to manipulate these arrays directly, however this is not recommended since one easily ends up with an ill-defined history resulting for example in segfaults or false results. Instead use the following methods to set the history.

hist_from_funcs *(dic, nn = 101)*

Initialise the histories with the functions stored in the dictionary dic. The keys are the variable names. The function will be called as f(t) for t in [-maxdelay, 0] on nn samples in the interval.

This function provides the simplest way to set the history. It is often convenient to use python lambda functions for f. This way you can define the history function in place.

If any variable names are missing in the dictionaries, the history of these variables is set to zero and a warning is printed. If the dictionary contains keys not matching any variables these entries are ignored and a warning is printed. Example: Initialise the history of the variables x and y with cos and sin functions using a finer sampling resolution:

```python
from math import sin, cos

histdic = {
    'x': lambda t: cos(0.2*t),
    'y': lambda t: sin(0.2*t)
}

dde.hist_from_funcs(histdic, 500)
```

hist_from_arrays *(dic, useend = True)*

Initialise the history using a dictionary of arrays with variable names as keys. Additionally a time array can be given corresponding to the key t. All arrays in dic have to have the same lengths.

If an array for t is given the history is interpreted as points (t,var). Otherwise the arrays will be evenly spaced out over the interval [-maxdelay, 0].

If useend is True the time array is shifted such that the end time is zero. This is useful if you want to use the result of a previous simulation as the history.

If any variable names are missing in the dictionaries, the history of these variables is set to zero and a warning is printed. If the dictionary contains keys not matching any variables (or 't') these entries are ignored and a warning is printed.

Example:

```
t = numpy.linspace(0, 1, 500)
x = numpy.cos(0.2*t)
y = numpy.sin(0.2*t)

histdic = {
    't': t,
    'x': x,
    'y': y
}
dde.hist_from_arrays(histdic)
```

Note that the previously used methods `hist_from_dict`, `hist_from_array` and `hist_from_func` (the last two without s) have been removed, since it was too easy to make mistakes with them.

16.2.3 The Solution

After the solver has run, the solution (including the history) is stored in the dictionary `dde23.sol`. The keys are again the variable names and the time `'t'`. Since the solver uses an adaptive step size method, the solution is not sampled at regular times.

To sample the solutions at regular (or other custom spaced) times there are two functions.

`sample` (*tstart = None, tfinal = None, dt = None*)
Sample the solution with `dt` steps between `tstart` and `tfinal`.
`tstart`, `tfinal` Start and end value of the interval to sample. If nothing is specified `tstart` is set to zero and `tfinal` is set to the simulation end time.
`dt` Sampling size used. If nothing is specified a reasonable value is calculated.
Returns a dictionary with the sampled arrays. The keys are the variable names.
The key `'t'` corresponds to the sampling times.

`sol_spl`(*t*)
Sample the solutions at times `t`.
`t` Array of time points on which to sample the solution.
Returns a dictionary with the sampled arrays. The keys are the variable names.
The key `'t'` corresponds to the sampling times.
These functions use a cubic spline interpolation of the solution data.

16.2.4 Noise

Noise can be included in the simulations. Note however, that the method used is quite crude (an Euler method will be added which is better suited for noise dominated dynamics). The deterministic terms are calculated with the usual Runge-Kutta method and then the noise term is added with the proper scaling of $\sqrt{dt}$ at the final step. To get accurate results one should use small time steps, i.e., dtmax should be set small enough.

The noise is defined in a separate dictionary. The function gwn() can be accessed in the noise string and is a Gaussian white noise term of unit variance. The following code specifies an Ornstein-Uhlenbeck process:

```
eqns = { 'x': '-x' }
noise = { 'x': 'D * gwn()'}
params = { 'D': 0.00001 }
```

```
dde = dde23(eqns=eqns, params=params, noise=noise)
```

You can also use noise terms of other forms by specifying an appropriate C-function (see the section on custom C-code).

16.2.5 Custom C-Code

You can access custom C-functions in your equations by adding the definition as supportcode for the solver. In the following example a function f(w, t) is defined through C-code and accessed in the eqn string:

```
# define the eqn f is the C-function defined below
eqns = { 'x': '- x + k*x(t-tau) + A*f(w,t)' }
params = {
    'k'  : 0.1,
    'w'  : 2.0,
    'A'  : 0.5,
    'tau': 10.0
}

mycode = """
double f(double w, double t) {
    return sin(w * t);
}
"""
```

```
dde = dde23(eqns=eqns, params=params, supportcode=mycode)
```

When defining custom code you have to be careful with the types. The type of complex variables in the C-code is Complex. Note in the above example that w has to be given as an input to the function, because the parameters can only be accessed from the eqns string and not inside the supportcode.

Using custom C-code is often useful for switching terms on and off. For example the Heaviside function may be defined and used as follows:

```
# define the eqn f is the C-function defined below
eqns = { 'z:c': '(la+ii*w)*z - Heavi(t-t0)* K*(z-z(t-tau))' }
params = {
    'K'  : 0.1 ,
    'w'  : 1.0,
    'la' : 0.1,
    'tau': pi,
    't0' : 2*pi
}


mycode = """
double Heavi(double t) {
    if(t>=0)
        return 1.0;
    else
        return 0.0;
}
"""
dde = dde23(eqns=eqns, params=params, supportcode=mycode)
```

This code would switch a control term on when t>t0. Note that Heavi (t-t0) does not get translated to a delay term, because Heavi is not a system variable. Since this scenario occurs so frequent the Heaviside function (as defined above) is included by default in the source code.

16.2.6 Use and Modify Generated Code

The compilation of the generated code is done with scipy.weave. Instead of using weave to run the code you can directly access the generated code via the function dde23.output_ccode(). This function returns the generated code as a string which you can then store in a source file.

To run the generated code manually you have to set the precompiler flag #define MANUAL (uncomment the line in the source file) to exclude the python related parts and include some other parts making the code a valid stand alone

source file. After this the code should compile with g++ -lm -o prog
source.cpp and you can run the program manually.

You can specify the history of all variables in the source file by setting the for
loops after the comment\ /* set the history here ... */.
Running the code manually can help you debug, if some problem occurs and also
allows you to extend the code easily.

16.2.7 Another Example

The following example shows some of the things discussed above. The code
simulates the Lang-Kobayashi laser equations [5] (see Chap. 11)

$$E'(t) = \frac{1}{2}(1 + i\alpha)nE + KE(t - \tau),$$

$$Tn'(t) = p - n - (1 + n)|E|^2.$$

```
import numpy as np
import pylab as pl
from pydelay import dde23

tfinal = 10000
tau = 1000
#the laser equations
eqns = {
      'E:c': '0.5*(1.0+ii*a)*E*n + K*E(t-tau)',
      'n'  : '(p - n - (1.0 +n) * pow(abs(E),2))/T'
}

params = {
      'a'   : 4.0,
      'p'   : 1.0,
      'T'   : 200.0,
      'K'   : 0.1,
      'tau' : tau,
      'nu'  : 10**-5,
      'n0'  : 10.0
}

noise = { 'E': 'sqrt(0.5*nu*(n+n0)) * (gwn() + ii*gwn())' }

dde = dde23(eqns=eqns, params=params, noise=noise)
dde.set_sim_params(tfinal=tfinal)
```

```
# use a dictionary to set the history
thist = np.linspace(0, tau, tfinal)
Ehist = np.zeros(len(thist))+1.0
nhist = np.zeros(len(thist))-0.2
dic = {'t' : thist, 'E': Ehist, 'n': nhist}

# 'useend' is True by default in hist_from_dict and thus the
# time array is shifted correctly
dde.hist_from_arrays(dic)

dde.run()

t = dde.sol['t']
E = dde.sol['E']
n = dde.sol['n']
spl = dde.sample(-tau, tfinal, 0.1)

pl.plot(t[:-1], t[1:] - t[:-1], '0.8', label='step size')
pl.plot(spl['t'], abs(spl['E']), 'g', label='sampled solution')
pl.plot(t, abs(E), '.', label='calculated points')
pl.legend()

pl.xlabel('$t$')
pl.ylabel('$|E|$')

pl.xlim((0.95*tfinal, tfinal))
pl.ylim((0,3))
pl.show()
```

Figure 16.2 shows the resulting plot.

Fig. 16.2 Numerical
solution of the LK equations

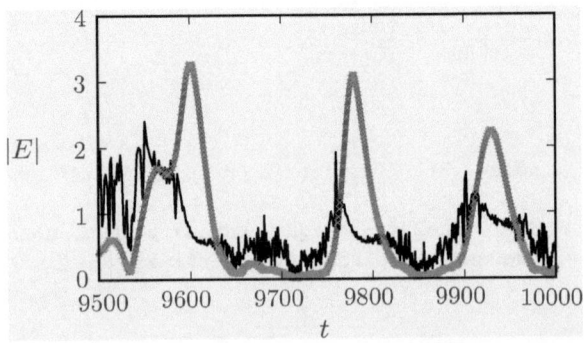

16.3 Module Reference

__init__(*eqns, params = None, noise = None, supportcode =", debug = False*)
Initialise the solver.

eqns Dictionary defining for each variable the derivative. Delays are written as
as (t-...) example:

```
eqns = {
    'y1': '- y1 * y2(t-tau1) + y2(t-tau2)',
    'y2': 'a * y1 * y2(t-tau1) - y2',
    'y3': 'y2 - y2(t-tau2)'
    }
```

You can also directly use numbers or combination of parameters as delays:

```
eqns = {
    'x1': '-a*x1 + x1(t - 1.0)',
    'x2': 'x2-b*x1(t-2.0*a+b)'
    }
```

At the moment only constant delays are supported.
The string defining the equation has to be a valid C expression, i.e., use
pow(a,b) instead of a**b etc. (this might change in the future):

```
eqns = {'y': '-2.0 * sin(t) * pow(y(t-tau), 2)'}
```

Complex variable can be defined using :C or :c in the variable name. The
imaginary unit can be used through ii in the equations:

```
eqns = {'z:C': '(-la + ii * w0) * z' }
```

params Dictionary defining the parameters (including delays) used in eqns. example:

```
params = {
    'a'   : 1.0,
    'tau1': 1.0,
    'tau2': 10.0
    }
```

noise Dictionary for noise terms. The function gwn() can be accessed in the noise string and provides a Gaussian white noise term of unit variance. Example:

```
noise = {'x': '0.01*gwn()'}
```

debug If set to True the solver gives verbose output to stdout while running.
set_sim_params($tfinal = None, tspan = None, AbsTol = 1e - 06$,
$\qquad RelTol = 0.001, dtmin = 1e - 06, dtmax = None, dt0 = None$,
$\qquad MaxIter = 1e9$)

tfinal End time of the simulation (the simulation always starts at t=0).

AbsTol, RelTol The relative and absolute error tolerance. If the estimated error e for a variable y obeys e <= AbsTol + RelTol*|y| then the step is accepted. Otherwise the step will be repeated with a smaller step size.

dtmin, dtmax Minimum and maximum step size used.

dt0 initial step size

MaxIter maximum number of steps. The simulation stops if this is reached.

hist_from_arrays ($dic, useend = True$)

Initialise the history using a dictionary of arrays with variable names as keys. Additionally a time array can be given corresponding to the key t. All arrays in dic have to have the same lengths.

If an array for t is given the history is interpreted as points (t, var). Otherwise the arrays will be evenly spaced out over the interval [-maxdelay, 0].

If useend is True the time array is shifted such that the end time is zero. This is useful if you want to use the result of a previous simulation as the history.

If any variable names are missing in the dictionaries, the history of these variables is set to zero and a warning is printed. If the dictionary contains keys not matching any variables (or 't') these entries are ignored and a warning is printed.

Example:

```
t = numpy.linspace(0, 1, 500)
x = numpy.cos(0.2*t)
y = numpy.sin(0.2*t)

histdic = {
    't': t,
    'x': x,
    'y': y
}
dde.hist_from_arrays(histdic)
```

`hist_from_funcs`(*dic*, *nn* = 101)

   Initialise the histories with the functions stored in the dictionary `dic`. The keys
   are the variable names. The function will be called as `f(t)` for `t` in
   `[-maxdelay, 0]` on `nn` samples in the interval.

   This function provides the simplest way to set the history. It is often convenient
   to use python `lambda` functions for `f`. This way you can define the history
   function in place.

   If any variable names are missing in the dictionaries, the history of these
   variables is set to zero and a warning is printed. If the dictionary contains keys
   not matching any variables these entries are ignored and a warning is printed.
   Example: Initialise the history of the variables `x` and `y` with `cos` and `sin`
   functions using a finer sampling resolution:

```python
from math import sin, cos

histdic = {
    'x': lambda t: cos(0.2*t),
    'y': lambda t: sin(0.2*t)
}

dde.hist_from_funcs(histdic, 500)
```

`output_ccode`(*manual* = *True*)

   Returns the simulation code as a string. If `manual` is True the generated code
   can be compiled and executed manually with minimal modifications.
   If `manual` is False the scipy-compatible code is returned.

`run()`

   run the simulation

class `dde23`(*eqns*, *params* = *None*, *noise* = *None*, *supportcode* = ", *debug* = *False*)

   This class translates a DDE to C and solves it using the Bogacki-Shampine
   method.

   `Attributes of class instances:`

   For user relevant attributes:

`self.sol` Dictionary storing the solution (when the simulation has finished).
   The keys are the variable names and `'t'` corresponding to the sampled times.

`self.discont` List of discontinuity times. This is generated from the occurring
   delays by propagating the discontinuity at `t=0`. The solver will step on these
   discontinuities. If you want the solver to step onto certain times they can be
   inserted here.

`self.rseed` Can be set to initialise the random number generator with a specific seed. If not set it is initialised with the time.

`self.hist` Dictionary with the history. Don't manipulate the history arrays directly! Use the provided functions to set the history.

`self.Vhist` Dictionary with the time derivatives of the history.

For user less relevant attributes:

`self.delays` List of the delays occurring in the equations.

`self.chunk` When arrays become to small they are grown by this number.

`self.spline_tck` Dictionary which stores the tck spline representation of the solutions. (see `scipy.interpolate`)

`self.eqns` Stores the eqn dictionary.

`self.params` Stores the parameter dictionary.

`self.simul` Dictionary of the simulation parameters.

`self.noise` Stores the noise dictionary.

`self.debug` Stores the debug flag.

`self.delayhashs` List of hashs for each delay (this is used in the generated C-code).

`self.vars` List of variables extracted from the eqn dictionary keys.

`self.types` Dictionary of C-type names of each variable.

`self.nptypes` Dictionary of numpy-type names of each variable.

References

1. V. Flunkert, E. Schöll, pydelay—a python tool for solving delay differential equations. arXiv:0911.1633 [nlin.CD] (2009)
2. P. Bogacki, L.F. Shampine, A 3(2) pair of Runge–Kutta formulas. Appl. Math. Lett. **2**, 321 (1989)
3. L.F. Shampine, S. Thompson, Solving DDEs in Matlab. Appl. Num. Math. **37**, 441 (2001)
4. M.C. Mackey, L. Glass, Oscillation and chaos in physiological control systems. Science **197**, 287 (1977)
5. R. Lang, K. Kobayashi, External optical feedback effects on semiconductor injection laser properties. IEEE J. Quantum Electron **16**, 347 (1980)

Index

V. Flunkert, *Delay-Coupled Complex Systems*, Springer Theses,   179
DOI: 10.1007/978-3-642-20250-6, © Springer-Verlag Berlin Heidelberg 2011